Buddhism and Intelligent Technology

Also available from Bloomsbury

A Critical Introduction to the Ethics of Abortion, by Bernie Cantens
An Ethical Guidebook to the Zombie Apocalypse, by Bryan Hall
Great Philosophical Objections to Artificial Intelligence, by Eric Dietrich, Chris Fields, John P. Sullins, Bram Van Heuveln and Robin Zebrowski
How to Think about the Climate Crisis, by Graham Parkes
Morality and Ethics at War, by Deane-Peter Baker
Nonviolent Resistance as a Philosophy of Life, by Ramin Jahanbegloo

Buddhism and Intelligent Technology

Toward a More Humane Future

Peter D. Hershock

BLOOMSBURY ACADEMIC
LONDON • NEW YORK • OXFORD • NEW DELHI • SYDNEY

BLOOMSBURY ACADEMIC
Bloomsbury Publishing Plc
50 Bedford Square, London, WC1B 3DP, UK
1385 Broadway, New York, NY 10018, USA
29 Earlsfort Terrace, Dublin 2, Ireland

BLOOMSBURY, BLOOMSBURY ACADEMIC and the Diana logo are trademarks of Bloomsbury Publishing Plc

First published in Great Britain 2021

Copyright © Peter D. Hershock, 2021

Peter D. Hershock has asserted his right under the Copyright, Designs and Patents Act, 1988, to be identified as Author of this work.

For legal purposes the Acknowledgments on p. x constitute an extension of this copyright page.

Cover image: © kjpargeter

All rights reserved. No part of this publication may be reproduced or transmitted in any form or by any means, electronic or mechanical, including photocopying, recording, or any information storage or retrieval system, without prior permission in writing from the publishers.

Bloomsbury Publishing Plc does not have any control over, or responsibility for, any third-party websites referred to or in this book. All internet addresses given in this book were correct at the time of going to press. The author and publisher regret any inconvenience caused if addresses have changed or sites have ceased to exist, but can accept no responsibility for any such changes.

A catalogue record for this book is available from the British Library.

A catalog record for this book is available from the Library of Congress.

ISBN: HB: 978-1-3501-8226-4
PB: 978-1-3501-8227-1
ePDF: 978-1-3501-8228-8
eBook: 978-1-3501-8229-5

Typeset by Newgen KnowledgeWorks Pvt. Ltd., Chennai, India

To find out more about our authors and books visit www.bloomsbury.com and sign up for our newsletters.

For my sons, Peter and Ka`eo.

Contents

Acknowledgments	x

Introduction		1
	Personal Presence, Ethics, and Global Action	7
	The Importance of Diversity for a Humanely Oriented Intelligence Revolution	12
	The Plan of the Book	15
1	**Buddhism: A Philosophical Repertoire**	19
	Buddhist Origins in Predicament Resolution	20
	Buddhist Practice: The Teaching of the Three Marks	29
	An Ethics of Compassionate Relational Virtuosity	37
2	**Artificial Intelligence: A Brief History**	43
	Servants of Our Own Making: Dreams of Artificial Beings and Mechanizing Reason	43
	Modeling Thought: The Research Origins of the Intelligence Revolution	45
	The AI Investment Winter and Its Aftermath	48
	A New Informational Infrastructure: The Internet, Personal Computer, and Smartphone	50
	Artificial Agency and the Goal of Intentional Partnership	52
	The Fourth Industrial Revolution: A Revolution in the Cloud	56
3	**Intelligent Technology: A Revolution in the Making**	61
	Intelligence: A Working Definition	61
	The Nonidentity of Tools and Technologies: A Critical Wedge	64
	Intelligence Industries and the New Attention Economy	67
	Commercial Intelligence Interests	73
	Political Intelligence Interests	80
	Military Intelligence Interests	88

	The Theater of Competition in the New Great Game: Consciousness	93
4	**Total Attention Capture and Control: A Future to Avoid**	**97**
	The Predicament of Wish-Fulfillment: The Midas Touch and Errant AI	98
	Contrasting Scenarios for Futures Free from Want	101
	The Future from Which Our Present Might Be Remembered	103
	Smart Services and the Sociopersonal Risk of Forfeiting Intelligent Human Practices	121
	The Predicament Ahead	125
5	**Anticipating an Ethics of Intelligence**	**127**
	Ethical Possibilities	128
	Ethical Resolution	144
6	**Dimensions of Ethical Agency: Confucian Conduct, Socratic Reasoning, and Buddhist Consciousness**	**151**
	The Ideal of Confucian Personhood: Resolutely Harmonious Conduct	153
	The Ideal of Socratic Personhood: Critical Resolve	158
	Blending Relational Intimacy and Rational Integrity	162
	Attentive Virtuosity: The Heart of Buddhist Ethical Effort	164
7	**Humane Becoming: Cultivating Responsive Virtuosity**	**169**
	Two Ideals of Personal Presence	169
	Nourishing Virtuosic Presence: The Six *Pāramitās*	176
	Personal Presence and Ethical Diversity in the Intelligence Revolution	187
8	**Course Correction: A Middle Path Through Data Governance**	**191**
	Digital Asceticism: A Way Forward?	193
	Digital Hedonism: An Impoverishing Alternative	194
	A Digital Middle Path: Resistance for Redirection	197
	Connectivity Strikes: Leveraging Citizen and Consumer Data Strengths	201

Data as the "Carbon" of the Attention Economy: Toward Public Good Data Governance	205
9 Course Correction: A Middle Path Through Education	**217**
Improvisational Readiness: The Essential Imperative of Twenty-First-Century Education	218
The First and Last Question: What Comes Next?	234
Notes	237
References	245
Index	255

Acknowledgments

The initial research for this book was supported by a nine-month fellowship with the Berggruen Institute. Through the fellowship I was able to spend six months in China and benefitted greatly from a number of the programs organized by the Berggruen Institute's China Center in Qufu, Beijing, and Shanghai. Conversations with Professor Liu Zhe at Peking University helped greatly in seeing Chinese policy on artificial intelligence from a thoughtfully enthusiastic and critical insider's perspective. Conversations with Liu Dele afforded insights not only into the entrepreneurial world in China but also into the spread of Buddhist practice in an increasingly digitalized Chinese society. Professor Roger T. Ames, also at Peking University and emeritus at the University of Hawai'i, was as always an inspiring embodiment of Confucian commitments to harmonious relationality. Finally, much gratitude is due to the Yun Hwa Denomination of World Social Buddhism and its founder, Ji Kwang Dae Poep Sa Nim, without whose exemplification of superlatively embodied compassion and clarity the presentation of Buddhism in this book would not have been possible.

Introduction

We live in revolutionary times. Exponential increases in the power and scope of artificial intelligence (AI) are being realized through the combination of big data and self-improving machine learning algorithms, and the technological and industrial promise is seemingly limitless. With good reason, many people are heralding this as the onset of the Fourth Industrial Revolution. Yet, this new revolution is as metaphysical as it is industrial. Like the Copernican Revolution, which radically decentered humanity in the cosmos five centuries ago, the data- and attention-driven Intelligence Revolution is dissolving once-foundational certainties and opening entirely new realms of opportunity.

Smart cities will be both more efficient and more livable. Smart healthcare has the potential to reach and benefit the half of humanity that now lacks even basic health services. Smart schools will respond to individual student needs, drawing on data generated by millions of teachers and learners worldwide. Yet, it is estimated that over the next twenty years as many as half of all core job tasks are liable to being taken over by artificial agents, even in such white collar professions as law and medicine. The digital economy is proving to be structurally biased toward monopoly. The infrastructure that enables continuous social media connectivity also supports corporate data mining and state surveillance. And smart services and the algorithmic tailoring of experience have potentials for dramatically transforming the meanings of family, friendship, health, work, security, and agency, at first supplementing and then eventually supplanting intelligent human practices. The immense technological promises of the Intelligence Revolution are kin to equally immense perils. This book is an attempt to bring into clearer focus the complex interplay of the technological, societal, and personal dimensions of the Intelligence Revolution and to encourage sustained intercultural and intergenerational ethical collaboration in figuring out how best to align AI with globally shared human values.

In recent years, considerable attention has been given to both the welcome and the worrying prospects of realizing artificial general intelligence or superintelligence. For techno-idealists, the advent of artificial general intelligence will mark an evolutionary leap from carbon to silicon and will bring an exponential expansion of intelligence and the freeing of consciousness from the constraints of biology. Techno-realists see, instead, potentials for an exponential scaling up of unintended consequences and risks, including threats to the continued existence of humanity. In both cases, attention is focused on the possibility of a *technological singularity*: a historical juncture at which technology acquires infinite value and artificial agents are free either to act upon or ignore human aims and interests.

Estimates regarding the creation of artificial general intelligence vary widely, from a mere quarter century to several centuries, and our chance of arriving at a technological singularity is thus very much a matter of speculation. What is not a matter of speculation—and what should concern us no less profoundly and much more immediately—is the *ethical singularity* toward which we are being hastened by machine agents acting on the intentions of their human creators: a point at which evaluating competing value systems and conceptions of humane intelligence take on infinite value/significance.

Ethics can be variously defined. But, at a minimum, it involves going beyond using our collective human intelligence to more effectively reach existing aims and using it instead to discriminate qualitatively among both our aims and our means for realizing them. In short, ethics is the art of human course correction. The ethical singularity ahead is the point at which the opportunity space for further human course correction collapses—a historical chokepoint at which we will have no greater chance of escaping the consequences of scaling up our often conflicting values than light has of escaping a cosmological black hole.

Navigating through and beyond this ethical singularity toward more humane and equitable futures will require exercising new kinds of collaborative ethical agency. For the first time, the shape of the human-technology-world relationship is not being wrought solely by human hands. Technology is no longer a passive medium through which humanity rearticulates what it means to be human and redefines the world of human experience. Technology is now intelligent: an active and adaptive participant in the human-technology-world relationship. Machine autonomy today is nowhere close to being on par with human autonomy, and may never be so. But machine agency is evolving with ever-increasing speed and it is doing so in ways that often cannot be fully explained. The ethical challenges posed by this new relationship are unprecedented.

Engineering machine evolution can be intoxicating. At gatherings of the scientists, technologists, engineers, and entrepreneurs working at the cutting edges of AI research and development, the excitement is almost palpably radiant. These are brilliantly curious people committed to answering the intellectual and practical challenges of opening, exploring, and exploiting vastly new realms of human and technological possibility. They are visionaries engaged in turning dreams into realities. For them, bringing ethics to bear on the development of AI is often understood as a matter of computationally achieving acceptable levels of transparency and accountability, assuring compliance with existing laws and regulatory frameworks, and ensuring that machine intelligences demonstrate operational/behavioral respect for accepted societal norms. Discussions tend to focus on developing the tools and techniques needed to achieve these goals and on how best to carry on with the exciting technical and scientific work ahead.

At gatherings of more skeptically inclined philosophers, legal scholars, and social scientists, the atmosphere is much more sober and energies are often directed toward articulating concerns about the moral hazards of adventuring in the realms of possibility that are being opened by the new intelligence industries. In keeping with the principle that is it better to be safe than sorry, these historically cognizant, and often profoundly thoughtful people would have us attend to actual and potential perils concealed in the shadows of technological promise. The most cautious among them urge immediate and rigorous consideration of the existential threat that the advent of artificial superintelligence might pose for humanity. In these gatherings, bringing ethics to bear on the development of AI and machine agencies begins with acknowledging that, good intentions notwithstanding, our technological achievements have often affected the human experience in ways that are malign as well as benign. Discussions tend to focus on preemptive caution, on strategies for slowing the pace of technological change so that we don't find ourselves spun out of control, and on establishing protocols for ensuring that AI, machine learning, and big data are developed and deployed in ways that are aligned with human values and are conducive to both personal and societal well-being.

Both kinds of discussions are necessary and valuable. But we will need something more if we are going to resolve the global predicament into which the Intelligence Revolution is conveying us. If there is one great lesson to be learned from the history of the human-technology-world relationship, it is that new technologies have both anticipated and unanticipated consequences and it is the unintended, ironic consequences which most dramatically alter the human and world dimensions of the relationship. If there is a parallel lesson to be learned from the history of ethics, it is that even the most ardently reasoned claims about universal human values and our options for course correction are historically and culturally conditioned. Significant gulfs already exist, for example, among the approaches to "ethical" or "human-centered" AI that are being taken in the United States, China, and the European Union, including sharply opposing perspectives on the importance of data privacy and net neutrality, and on how to weight individual, corporate, and state interests in efforts to align AI with human values and societal well-being. The path toward global ethical consensus on intelligent technology is far from apparent.

It is tempting to assume that the promises and perils of now-emerging intelligence technologies and industries can be clearly distinguished, and to conclude that courses of action enabling us to avoid the latter and happily realize the former can be readily identified in advance. Unfortunately, there is no factual warrant for this assumption. On the contrary, the greatest perils of intelligent technology do not appear to be extrinsic to its great promises but rather intrinsic to them. The rise of autonomous vehicles will lower risks of traffic accidents and reduce insurance burdens. But, in doing so, it will also curtail the employment of heavy equipment operators and truck and taxi drivers. Virtual personal assistants will provide smart services drawing on knowledge resources far greater and far more quickly than any human ever could but will at the same time subject their users to profit-generating surveillance and contribute to the atrophy of an increasingly wide range of intelligent human practices.

Thinking of such perils as mere side-effects of realizing the promises of the Intelligence Revolution is both practically and ethically misleading. Side-effects are contingencies that can be controlled for and, in most cases, bypassed on our way to achieving desired ends. They are obstacles that can be gotten around by means of new techniques or practices that enable us to continue forward without making any significant changes in our motivating values, aims, and interests. But in much the same way that mass-produced manufactured goods and climate change are equally primary products of industrialization based on fossil fuel burning when conducted at global scales, the promises and perils of the Intelligence Revolution are analogously primary products of global intelligence industries aimed (as they currently are) at reaping the benefits of machine learning and big data.

The analogy is instructive. Climate change is not a *problem* awaiting technical solution. It a *predicament* that makes evident conflicts within and among prevailing constellations of social, cultural, economic, and political values. Our collective failure to generate the commitments and collaborative action needed to alleviate and eliminate climate change impacts is thus not a technical failure; it is an ethical failure.

Unlike problems, predicaments cannot be solved for the simple reason that the value conflicts that they express make it impossible to determine exactly what would count *as* a solution. Climate change could easily be slowed or stopped by dramatically cutting back and eventually eliminating carbon emissions. What prevents this are conflicts among environmental values and those that undergird desires for rapid economic growth, political stability, and cultural continuity. Predicaments can only be resolved, where resolution implies both clarity and commitment—*clarity* about how an experienced value conflict has come to be and *commitment* to evaluating and appropriately reconfiguring relevant constellations of values, aims, and practices. Predicament resolution is thus inherently reflexive. It involves changing not only *how* we live but both *why* and *as whom*.

Much like climate change, the Intelligence Revolution is forcing confrontation with the value conflicts that underlie its pairing of benign and malign consequences. But the predicament into which humanity is being enticed by the Intelligence Revolution is particularly complex. Systems of AI are functioning as agents of experiential and relational transformation, actively altering the humanity-technology-world relationship in accord with computational directives to reinforce the readiness of ever more precisely desire-defined individuals to accept ever-greater connective convenience and choice in exchange for granting corporations and states new and evolving powers of control. Historically unprecedented, these systems operate according to a new logic of domination that is not expressed in overt acts of *coercion* but through soliciting voluntary membership in network-enabled regimes of ambiently reinforced *craving*. It is freely spent human attention energy and the data—the traces of human intelligence—that are carried along with it which are being used by competing corporate and state actors to build smart societies, to individually tailor human experience and behavior, and to incentivize reliance on smart services in ways that have the potential to render human intelligence superfluous.

The value tensions made evident by the Intelligence Revolution are at one level intimately personal—tensions between connective freedom and privacy, for example. But they are also political. The Second Industrial Revolution, from the 1870s to the 1940s, was a key driver for scaling up imperial and colonial ambitions in the so-called Great Game played for global control over land and labor. Similarly, the Intelligence Revolution—the Fourth Industrial Revolution—is driving a New Great Game: a competition among commercial, state, and nonstate actors seeking digital control over human attention and, ultimately, dominance in nothing less than the colonization of consciousness itself.

The technologies of the first three industrial revolutions—epitomized by the steam engine, scientific mass production, and digitalized computing and communications—scaled up human intentions for over two hundred years before it became evident that continued industrial uses of fossil fuels would irreversibly alter the planetary climate system. Our inability to resolve the climate predicament and halt or reverse climate impacts over the fifty years since then give us little cause for optimism in the face of the predicament posed by intelligent technology. The Fourth Industrial Revolution is likely to scale up human intentions and value conflicts to the point of irreversible changes in human presence and the dynamics of the anthrosphere over a matter of one or two decades, not one or two centuries.

Given the disparate approaches to "ethical" or "human-centered" AI that are being taken, for example, in the United States, the European Union and China, it's tempting to frame the New Great Game as a competition among different national or regional visions of the "good life" in a "smart society." But this way of seeing things directs critical concern exclusively toward which player(s) could or should win, rather than toward the human/societal impacts of playing the game, regardless of who ends up "winning." Geopolitics is important. Concerns about an AI "arms race" are very real. And the process of establishing global frameworks for ethical AI will undoubtedly be one of craftily negotiated steps forward, backward, sideways, and then forward again. But, framing the New Great Game as a two-dimensional conflict between, for instance, a choice-valuing "West" and a control-valuing "East" cannot do justice to either its geopolitical or its ethical complexity.

Contrary to our historically grounded intuitions, the predicament at the heart of the Intelligence Revolution is inseparable from the fact that the new digital systems of domination are becoming ever more intimately personal the more global their structures of data gathering and machine learning become. This fusion of the personal and the global is not incidental. The Intelligence Revolution would have been impossible without the connectivity explosion that began with the smartphone and global wireless internet access, and that is now being hypercharged by exponential growth in the internet of things. It was the planet-engulfing flood of data released by 24/7 connectivity through portable and increasingly miniaturized wireless devices that made it possible for machine learning to become something other than a tech-lab curiosity. Informed by ever greater volumes, varieties, and velocities of data flows, including data about our day-to-day lives, our loves and our longings, machine learning capabilities have evolved with breathtaking speed, penetrating and increasingly pervading virtually every domain of human endeavor. The personal and the global are being brought into algorithmic communion.

In addition, with nearly seamless worldwide connectivity and a mass transfer of social energy from physical to virtual spaces of interaction, not only is the divide between the private and the public eroding, the personal is becoming nearly indistinguishable from the interpersonal. At the same time that the geographical coordinates of our bodily presences are becoming globally accessible, our social, political, and economic presences are becoming increasingly delocalized. The Cartesian conviction that "I think, therefore I am" is giving way to a new metaphysics of personal presence summed up in the realization that "as we connect, so we are."

The power implications of this shift are profound. It is already apparent that the search and recommendation algorithms that mediate global connectivity are wielding immense epistemic power, actively shaping informational currents and thus what we encounter and think about and know. The resulting concentration of economic power in key platform and social media providers over the last fifteen years has been nothing short of stunning. Yet, the power to shape connectivity is also ontological power—the power to shape both who we are and who we become. Recognizing this is the first step toward engaging the Intelligence Revolution, not as a technological inevitability but as an ethical challenge and opportunity. Central to the task of meeting this challenge will be envisioning who we need to be *present as* if we are going participate in the Intelligence Revolution as an opportunity for recentering humanity, improvising together in shared commitment to equitable and truly humane futures.

Personal Presence, Ethics, and Global Action

Asking who we need to be *present as* in order to resolve the predicament of AI might be interpreted as a question about how to achieve and sustain individual well-being in a time of dramatic technological change. That, however, is not what I have in mind. Self-help guides to life in a digitally intelligent world may be of great value individually. But the work of establishing more equitable and truly humane trajectories for the Intelligence Revolution will not be accomplished by simply going into individual retreat. Exercising our "exit rights" from social media, online shopping, and smart services might be necessary—at least temporarily—to be able to engage in this complex work. But it will not be sufficient. Opting out completely from the dynamics of network connectivity is, in effect, to opt out of being in a position to inflect or redirect those dynamics. Avoiding a predicament is not a viable means to resolving it.

Alternatively, emphasizing who we need to be *present as* to respond effectively to the ethical singularity toward which we are being impelled by the Intelligence Revolution might suggest a preemptive commitment to virtue ethics—an emphasis on cultivating the character traits needed to lead ethically sound lives and to deliberate effectively about what it might mean to engineer humane systems of artificial agency. Appeals to virtue ethics have, in fact, formed one of the major currents in recent attempts to frame ethics policies for research and development in AI and robotics. However, as will be discussed in Chapter Five, there are very good reasons to avoid relying exclusively on any single ethical tradition. Moreover, as resources for truly global predicament resolution, systems of virtue ethics have the liability of being built around culturally specific and often conservative commitments to virtues or character traits that are very much historically conditioned. The ethical deliberation involved in truly global predicament resolution must be both intercultural and improvisational.

A New Metaphysics: The Centrality of Interdependence

The aim of focusing on qualities of personal presence is to prepare ourselves for the personal challenges of intercultural and improvisational ethical deliberation. But the underlying reasons for stressing qualities of personal presence have to do with what we might call the everyday metaphysics and economics of the Intelligence Revolution, and in particular their erasure of clear boundaries between the global and the personal, between the public and the private, and between control and choice.

As was just intimated, one of the guiding premises of this book is that the most pressing challenges of the present and coming decades—climate change and the transformations associated with the Intelligence Revolution central among them, but including as well the degradation of both natural and urban environments; the persistence of global hunger; and rising inequalities of wealth, income, risk, and opportunity—cannot be responded to adequately and resolved sustainably unless we are able to grapple with and resolve the complex value conflicts that they objectively express. In the terms used earlier, we are in the midst of an era-defining shift from the dominance of problem solution to that of predicament resolution: a shift from the primacy of the technical to that of the ethical.

The problem-to-predicament shift, however, is also evidence of the end of an era in which changes and challenges could be met and managed in the context of a world that remained essentially familiar. Predicament resolution is not a matter of simply choosing among ends and interests like destinations listed on an airline departures board. It involves improvising holistic and ultimately irreversible shifts of existential direction: a process of responsive and responsible self-transformation that, because it alters the totality of our relational possibilities, is always also world-altering. Put in the language of social science, we are in the midst of an era-defining transmutation of the interplay of structure and agency.

As earlier alluded, this ongoing transmutation of the world is no less dramatic than that brought about by the Copernican Revolution some five hundred years ago. Copernicus was confronted with evidence that something was profoundly and fundamentally amiss in prevailing assumptions about the structure of the cosmos. There seemed no way to reconcile observations of the movements of celestial bodies like the stars, planets, and the Moon with the commonsense and apparently divinely ordained belief that the Earth was at the center of the universe. To realize, as Copernicus did, that the Earth orbited the Sun, rather than the other way around, meant realizing that our everyday experience of the Sun rising in the east and setting in the west is simply an illusion of perspective. This realization was as much metaphysical as it was physical, and it shattered a host of previously foundational certainties. Most immediately, these were intellectual and religious. But over time, the casualties also came to include long-standing social, political, and economic certainties. In this spreading absence of certainty, there opened entirely new spaces of opportunity.

Today, instead of a decentering of the Earth, we are undergoing a decentering of the individual: the dissolution of once-evident certainties about the primacy of personal independence and national sovereignty. It is becoming increasingly—and at times, painfully—apparent that continued belief in the primacy of the individual contributes to our being incapable of navigating the ethical straits we must pass through in order to resolve the global predicaments we face. Although we may persist in taking individual persons, religions, cultures, corporations, or nations as our points of reference and continue seeing global dynamics as revolving around them, we have strong evidence now that the centrality of the individual is an illusion. It is not the individual that is basic but the relational.

Just as daybreaks continue to appear to us as sun rising events rather than as Earth-turnings, centuries after Copernicus proved otherwise, most of us will "naturally" continue to experience ourselves as individuals. Hopefully, however, we will be as successful in acting on our new knowledge of the primacy of relationality as we have been in acting on our knowledge about the movements of celestial bodies. Acting on the knowledge that the Earth revolves around the Sun, we have sent probes to all of the solar system's planets and have even landed a robot surveyor on an asteroid the size of a few city blocks in an orbit between that of Mars and Jupiter. These are astonishing feats. Through the decentering of the individual, we are being introduced not only to entirely new kinds of challenges but also to entirely new kinds of opportunities. Once we begin acting consistently on our knowledge that the existence of individuals is a perspectival illusion, we will gain similarly astonishing new capabilities.

A New Economics: The Value of Attention

As long as the economics of the Intelligence Revolution remain on their current course, however, realizing these new capabilities equitably will not be easy. Although we continue to speak about a global information economy, big data has made this obsolete. Information is too cheap and abundant to serve either as real-world currency or as sought-after commodity. The core activity of emerging intelligence industries and their computational factories is processing the data circulating through global connectivity networks, discovering patterns in how we direct our attention, inferring from these patterns our personal values and interests, and then using these inferences to first anticipate and then shape human intentions. The core product of these industries and their factories is human behavior.

It is the systematic attraction and exploitation of attention that now drives global circulations of goods and services. In this new attention economy, revenue growth is directly proportional to the effectiveness of machine agencies in discerning what we value and predicting what we can be induced to want. It is an economy in which major platform, social media, and internet service providers are empowered to profit from algorithmically tailoring our increasingly connection-defined human experience. And the holy grail of this new economy—which I will call the Attention Economy 2.0—is not free energy or cheap labor. It is *total attention-share*.

We are, of course, being compensated for allowing our patterns of attention to be mapped and profitably modified. As consumers, in exchange for our attention, we are provided with individually tailored and yet seemingly infinite arrays of choice—for tangible goods and services, for the intangible experiences afforded by film and music, and for social connection. Yet, granted that sustained attention is the single indispensable resource required to engage in creative relational transformation, this is not a costless or value-neutral exchange. As machine and synthetic intelligences are fed ever more revealing access to real-time streams of behavioral, biometric, and other data, the algorithmic tailoring of digitally mediated experience will become ever more seamlessly effective in crafting experiential and connective destinies that are intimately expressive of our own yearnings and desires.

This might be considered by some of us as a dream come true: the technological provision of access to unique experiential paradises for each and every one of us. Others might consider it a nightmare in the making. As we will later see in greater detail, the immense commercial benefits of big data, machine learning, and AI depend on state-sanctioned perforations of the boundary between the public and the private—a marriage of the *attention economy* and the *surveillance state*. Like traditional arranged marriages that have the primary purpose of alloying and securing family fortunes, the technology-mediated marriage of commercial and state interests is not a love match. It is a highly calculated union through which a small number of global elites gain decisive (and for the most part invisible) control over where and how we connect, with whom, and for what purposes.

This concentration of power, and the funneling up of wealth and opportunity that accompany them, is cause enough for serious global worry. But the deeper concern—and what warrants seeing this already ongoing process of technological transformation as powerfully predicament-generating—is the fact that our ever greater and more individuated *privileges to choose* come at the cost of granting commercial and political actors ever more extensive *rights to control*. As noted earlier, this is not control via *coercion* but via *craving*. The technological route to paradise opened before us is one along which our experiential options will become both wider in scope and ever more acutely desirable but only to the extent that we trade away our "exit rights" from the connective—and thus relational—destinies that are being crafted for us with tireless machine ingenuity: a forfeiture, eventually, of experiential and relational wilderness in exchange for compulsively attractive digital captivity.

To be clear, the claim here is not that digital captivity is our inevitable human destiny. That is one direction—albeit a likely one at present—in which we might be carried by the digitally mediated fusion of dynamic global structures and individual agency. In fact, the rhetorical opposition of "wilderness" and "captivity" is no better in doing justice to the complexity of the transformations in which we find ourselves caught up than is the reduction of competitions in the New Great Game to a bipolar "Cold War" of algorithmic proliferation. Because of the intimate ways in which attention factors into these transformations and competitions, we are not caught up in them contingently; we are caught up both willingly and constitutively. We are captivated, at once enthralled and complicit together in the machine-enabled process of being remade.

This is not as pessimistic a claim as might first appear. The hope informing this book is that our involvement in this process will be both critical and caring. The "alternating current" of data flows and algorithmic responses that are streaming constantly through the intelligence-gathering infrastructure of global network connectivity are dissolving the validity conditions for claims about the opposition of free will and determinism. But even as keen attention to this process forces questioning the relationship of freedom and choice, it also invites creatively exploring the meanings of freedom beyond choice and of agency beyond individuality.

It is hard to fully grasp the fact that exercising greater freedoms of choice as individual agents is consistent with ever greater and more concentrated powers of structurally mediated control. This is especially true for those of us raised to see freedom of choice as a foundational value. The manipulative powers being acquired through the Intelligence Revolution are unprecedented in their precision and reach. In terms of extensive reach, these are powers to affect everything from consumption patterns to voting behavior. In terms of intensive reach, they are powers to affect everything from our emotional lives and dating behavior to our patterns of curiosity. Yet, our intimate implication in the process of generating these powers, and our awareness of the predicament of freedom in a world of "wish-fulfilling" technologies, also position us to play active roles in altering the dynamics of the new attention economy and shaping the course of the Intelligence Revolution.

It is admittedly tempting to argue that the scale and ambient nature of the systems involved render us individually powerless—incapable of personally determining how things turn out, even for ourselves or our families. Global, history-making forces are at work. But this apparent lack of individual *power* is nevertheless compatible with acquiring the relational *strength* needed to make a historical difference through collaborative deliberation and action. In the context of today's recursively evolving human-technology-world relationship, the difference between being implicated in a system of smart *manipulation* and being implicated in a system for intelligent and perhaps liberating *self-discipline* is ultimately not a technical matter of design. It is a profoundly ethical matter of attentive quality and values.

The Importance of Diversity for a Humanely Oriented Intelligence Revolution

A second guiding premise of this book is that global predicament resolution can only be carried out as an interpersonal, intercultural, and intergenerational process and that this implies the centrality of diversity as an ethical value and imperative. In other words, our possibilities for humanely redirecting the dynamics of the Intelligence Revolution are predicated on appreciating the necessity of concerted and resolutely coordinated global action in which differences are engaged as the basis for mutual contribution to sustainably shared flourishing. The book's pivotal question is: who do we need to be present as to engage globally in the work of shared predicament resolution?

One of the distinguishing features of the response offered will be my appeal to classical philosophy and, in particular, to Buddhist thought and personal ideals. Given the unprecedented nature of the ethical labor ahead, appealing to premodern philosophical traditions is perhaps counterintuitive. Our natural inclination is to turn to our own contemporaries for guidance. The logic for this is clear. Contemporary ethicists are able to take into account the nature of our present circumstances and challenges, formulating normative perspectives on personhood from within—and in direct response to—current patterns of technological mediation. Yet, as logical as it might be to appeal to our contemporaries for guidance, it can also be a liability.

If designing and implementing new technologies is simultaneously a process of designing and implementing new norms and processes for being (or becoming) human, the industrial and intelligence revolutions are as fundamentally processes of remaking ourselves as they are processes of crafting new intentional prostheses and strategic environments. Every human-technology-world system reveals new possibilities of presence and action while concealing others. As we will be exploring in detail, technologies are emergent systems of material and conceptual practices that embody and deploy both strategic and normative values: intentional environments that at once shape how we do things and why. As such, maturing technologies have the effect of naturalizing certain forms of agency and qualities of presence. Granted this, there is no escaping the fact that the ethical systems which have evolved in the specific contexts of modern histories of human-technology-world relations will (even if only indirectly) constrain our imaginations of who we can and should be as ethical agents.

The purpose of looking to premodern philosophical traditions is not to entertain the revival of a premodern way of life. We can no more revive past ways of life than we can return to our youths, nor should we wish to do so. Rather, the purpose is to make visible potentially valuable conceptions of exemplary human presence that the evolutionary history of the modern human-technology-world relation has written over in zealous expression of its own salvific inevitability. Carefully engaged, premodern, and indigenous traditions can afford us critical sanctuary. They open perspectives on thinking together toward shared futures not only from "before" the conceptual bifurcation of the individual/personal and the collective/social but also from "before" the technological marvel of an intelligence-gathering infrastructure that profitably fuses personalization and popularization, creating an immaterial alloy of uniqueness of choice and universality of connection that, when algorithmically sharpened, may prove capable of severing the roots of human responsibility and creativity.

Turning to Buddhism for critical insight regarding the risks and most apt responses to intelligent technology is, admittedly, an especially counterintuitive move. In part, it is simply a reflection of my own background. My doctoral training was in Asian and intercultural philosophy with a focus on Chinese Buddhist traditions, and I have maintained a daily Buddhist meditation practice for almost forty years. But there are three more substantive rationales for turning to Buddhist thought and practice for insight.

First, Buddhism was founded as a practical response to the conflicts, troubles and suffering that result when the interdependent origins of all things are ignored in attachment to individual, self-centered existence—a tradition of keen and critical attunement to relational qualities and dynamics, rather than to individual agents and their actions. As such, it offers distinctive conceptual resources for exploring both the experiential and structural ramifications and risks of intelligent technology. Second, as a tradition rooted practically in attention training, it opens prospects both for developing a much-needed, critical phenomenology of the attention economy and for engaging in disciplined and yet freedom-securing resistance to it. Finally, in ways that are particularly useful in understanding the risks of a human-technology-world relationship shaped by evolutionary algorithms creatively intent on actualizing conflicting human values, the Buddhist concept of karma uniquely highlights the inseparability of fact and value and thus the ultimate nonduality of metaphysics and ethics.

Given the popular association of karma with everything from stories of past lives and inescapable fates to bland truisms like "what goes around, comes around," it is perhaps useful to offer a preliminary characterization of the Buddhist teaching of karma. Stated as a conditional, the Buddhist teaching of karma is that if we pay sufficiently close and sustained attention to our own life experiences, it becomes evident how abiding patterns of our values, intentions, and actions invariably occasion consonant patterns of experienced outcomes and opportunities. That is, we discover that we live in a world that is irreducibly meaning laden—a world in which experienced realities always imply responsibility and in which what we conventionally objectify as matter is ultimately the definition of a point of view: an emergent function of what has mattered to us. Karmic ethics is thus an ethics of predicament resolution, an ethics concerned critically not only with how and why we think and act as we do but also with the qualities of consciousness and intention embodied in our thinking and acting. It is, in short, an ethics of commitment to realizing progressively liberating patterns of attention.

The Plan of the Book

Given the centrality of Buddhism to my interpretation of the Intelligence Revolution and my convictions about who we need to be present as to navigate through and beyond the ethical singularity it is precipitating, Chapter One offers a concise introduction to Buddhist thought. For readers unfamiliar with Buddhism, this will provide a body of considerations within which the various narrative threads about intelligent technology presented in the succeeding chapters will be able to productively resonate, much as the wooden body of an acoustic guitar amplifies and accentuates resonances when its strings are plucked or strummed.

One of the practical implications of a karma is that sustainably alleviating and/or eliminating conflict, trouble, and suffering is only possible on the basis of a clarity about how things have come to be as they are. That is, histories matter. Real possibilities for course correction in the context of the Intelligence Revolution depend on understanding, not only the factual historical confluences but also the aspirational narratives out of which intelligent technology has emerged. Chapter Two offers an overview of the history of AI that stresses key streams of research and development and their breakthrough mergers over the last decade.

Chapters Three opens by presenting working definitions of intelligence and technology and explores how intelligent technology is transforming the human lifeworld as corporate and state interests combine in the playing of the New Great Game for digital supremacy. Building on this structural overview of the Intelligence Revolution, Chapter Four looks prospectively forward toward the human-technology-world relationship that current applications of big data, machine learning, and AI seem likely to bring into being over the next ten to fifteen years. Chapter Five considers whether the existential risks, environmental threats, and inequalities being generated by the Intelligence Revolution can be dealt with adequately from within the horizons of existing ethical systems and then makes use of Buddhist resources to argue on behalf of the necessity of an intercultural or "ethical ecosystem" approach to addressing the global predicament of intelligent technology. Granted the soundness of that argument, the question of who we need to be present as to contribute to resolving the intelligence predicament can be restated more precisely as "who do we need to be present as to engage in diversity-enhancing ethical improvisation?"

Chapter Six offers a preliminary response to that question, exploring first the Confucian virtue of relational intimacy and the Socratic virtue of rational integrity and then their blending within the Buddhist ethical ideal of attentive and responsive virtuosity. Chapter Seven investigates Buddhist personal ideals of uncompelled and compassionately engaged presence and the centrality of improvisation in Buddhist ethics. The final two chapters reflect on how to move forward practically to open prospects for a humane turn in the Intelligence Revolution, realizing the conditions for turning aspirational ideals into realities employing a Middle Path strategy of resistance and redirection focused on data governance and education.

It is important to stress at the outset that the extensive use of Buddhist concepts to understand the dynamics and risks of the Intelligence Revolution, and to frame a set of personal ideals suited to engaging those dynamics and risks, is not meant to be exclusionary. Buddhist resources are very helpful in laying out not only why ethical improvisation is needed to resolve the predicament posed by intelligent technology but also why it is imperative to ecologically integrate ethical perspectives native to a wide range cultural traditions and historical periods. Rather than a Buddhist ethics of technology, what is presented in the pages to follow is a Buddhism-inflected rationale for the indispensability of ethical diversity in responding to the challenges of intelligent technology.

*

The transformation of the human-technology-world relationship by intelligent technology is not a blind process set in motion by transcendent forces, and what it means for our futures is our shared, human responsibility. Intelligent technology is exposing us to deepening structural risks of mass unemployment and underemployment; of precipitously deepening inequalities of wealth, income, and opportunity; and of outsourcing morally charged decision-making to autonomous machine agents. As currently oriented, the dynamics of intelligent technology are liable to fix in place and amplify conflicts among human values and intentions of the kind that are at the roots of global predicaments like climate change, the persistence of hunger in a world of food excess, and our collective human failure to realize conditions of dignity for all in an era of the greatest wealth generation in history. Machine intelligences are poised to diligently transform the world as servants, savants, soldiers, and solicitors scaling the best and worst of human intentions, mirroring back to us our own patterns of attention and inattention.

The Intelligence Revolution cannot carry humanity forward into a more humane and equitable world as long as *we* are incapable of consistently humane and equitable conduct. Realizing the liberating potentials of intelligent technology will depend on whether we succeed in liberating ourselves from our tendencies to embody conflicting values, intentions, and actions. Given humanity's historical track record, the prospects of realizing such liberating potentials might seem poor. But humanity has never been faced with such clear assurance of ultimate responsibility for the futures we share or such clear imperatives for shouldering that responsibility collaboratively. Given this, sober optimism is perhaps not a fruitless exercise in unreasonable or empty hope.

1

Buddhism: A Philosophical Repertoire

The history of Buddhism is roughly as long as the history of imagining intelligent artificial beings. But whereas the history of artificial intelligence (AI) as a scientific and technological quest begins in earnest only at the dawn of the modern era, Buddhist practices, institutions, and systems of thought were evolving with great vitality and with substantial social, political, and cultural impacts across South, Central, and East Asia as early as the second century. Most of this long history is of no particular relevance for who we need to be present as to engage in collaborative global predicament resolution. But some historical context is useful in appreciating the contemporary relevance of Buddhist personal ideals and their conceptual roots.

Although Buddhism is customarily presented as a "world religion," this is somewhat misleading. Prior to the invitation of Buddhist teachers from various parts of Asia to the 1893 World Parliament of Religions in Chicago, few Buddhists would have identified themselves generically as practitioners of "Buddhism." Instead, most Buddhists would have identified with one or more lines of transmission, passing through a specific teacher, text, or temple, that connected them to Buddhism's founding figure, Siddhartha Gautama, who is generally referred with the honorific title the "Enlightened One" or Buddha. Traveling across premodern Asia, one would have passed through dramatically differing "ecologies of enlightenment" thriving at scales ranging from the local to the regional, each one of which would have been characterized by a distinctive set of personally transmitted practices and supporting teachings.

The differences among Buddhist traditions are in part a function of their historical depth. Although scholars continue to debate the dates of the Buddha's life, Buddhist practices and teachings date back to at least the fifth century BCE. As these practices and teachings spread from their origin in the Himalayan foothills across most of Eurasia over the next twelve hundred years, substantial differentiation would be expected as a matter of course. But in addition—and in contrast with religions built on foundations of divine revelation—Buddhism valued adaptive variation. For example, while the earliest strata of teachings attributed to the historical Buddha have remained important to the present day, the Buddhist canon remained open for well over a thousand years after first being put into writing in the first century BCE.

This openness was quite self-conscious. Buddhism originated in the context of dramatic rural-to-urban migration and small-scale industrialization on the Indian subcontinent and spread along with expanding trade throughout South Asia and then into Central, East, and Southeast Asia. An important factor in the rapidity of this spread was the Buddha's insistence that his teachings be conveyed in local languages and in ways that were both accessible to different audiences and responsive to local concerns.

In addition, unlike religious traditions based on divinely revealed truths that must be accepted on faith, Buddhist teachings were presented as perspectives on the human experience that could be verified personally—instructions or guidelines for truing or better aligning our ways of being present. Instead of theoretical reflections or metaphysical declarations, Buddhist teachings laid out a therapeutic system for understanding and alleviating psychological and social malaise. In sum, Buddhism originated as a purposefully evolving repertoire of practices for personally realizing liberating forms of relationality.

Buddhist Origins in Predicament Resolution

It is significant that Buddhist teachings originated in response to the challenges of day-to-day life—including those of sickness, old age, and death—during a time of rapid societal change, mass migration, and expanding trade. As is true today in urban centers around the world, when people from different cultural and moral communities find themselves living in close company, values conflicts are practically unavoidable. Buddhist teachings thus responded, at least in part, to the personal and cultural predicaments that we experience when compelled to make conscious and often uncomfortable decisions about which of our customs and identities to abandon and which new customs and identities to adopt in their place. In such disconcerting contexts, we often discover that what we had assumed to be universal commonsense is not actually common to everyone and that many of our most deeply ingrained habits of thought, feeling, speech, and action are as likely to trigger discord as they are to bring expected results.

In its most succinct formulation, the Buddhist therapeutic system is built around just four "truths" or "realities" (*satya*): (1) the presence of conflict, trouble, and suffering (*dukkha*; *duḥkha*) in the human experience; (2) the origination and persistence of conflict, trouble, and suffering as a function of value-infused patterns of causes and conditions; (3) the presence of possibilities for disrupting and dissolving those patterns; and (4) the existence of a pathway or method for accomplishing this personally through the embodied realization of moral clarity (*śīla*), attentive mastery (*samādhi*), and wisdom (*paññā; prajñā*).[1] These four truths/realities are not transcendentally derived eternal declarations about the origins and purpose of the world or our places in it. They are points of embarkation for revising *how we are present*.

Interdependence: The Primacy of Relationality

The pivotal insight informing the Buddhist therapeutic system is that all things and beings arise interdependently (*paṭiccasamuppāda; pratītyasamutpāda*). Thanks to the environmental crises triggered by human industrial activity over the last half century, this claim is not as strikingly novel today as it was during the Buddha's lifetime. We are now at least intellectually familiar with the idea that everything in nature is deeply interconnected, but also that we are intimately interconnected with one another and with nature. The interdependence of economies around the world is now taken for granted, and awareness of the global predicament of human-induced climate change has made practical engagement with patterns of deep interconnection a transnational moral and political imperative.

At the time Siddhartha Gautama had this insight—the culmination of his six-year search to discover how to sustainably alleviate experiences of conflict, trouble, and suffering—it ran very much counter to all prevailing religious and philosophical convictions about the nature of reality. These convictions were arrayed along a spectrum. At one end were convictions about the ultimate reality of "spirit/mind." At the other end were convictions about the ultimate reality of "matter/body." In between was a shifting array of hybrid or dualist views. The so-called Middle Path forwarded by the Buddha ran perpendicular or oblique to this spectrum rather than taking up a position somewhere along it. In effect, it was a method for moving beyond reductions of reality to some primal 'this' or 'that', as well as beyond combinations thereof. In philosophical terms, the Middle Path denies validity both to metaphysical monism (reality ultimately consists in just one kind of thing) and to metaphysical pluralism (reality ultimately consists in many kinds of things). It is a path of realizing, both personally and progressively, that what is ultimately real is relationality.[2]

At one level, to see all things as interdependent is to see how causality is always in some degree mutual rather than linear or one way. Mutual causality is, of course, at the heart of the contemporary science of ecology, which explores interdependence as the basis of healthy biological organization in natural ecosystems. In ecosystems, each species—plant, animal, insect, and microbial—makes some special contribution to the vibrancy and resilience of the system as an emergent and relationally sustained whole. Each species both affects and is affected by all of the other species in the system.

This way of describing things suggests, however, that individual plants, animals, insects, and microbes are basically separate entities that have various kinds of contingent relationships with one another. It suggests, for instance, that a tiger captured in the wild and relocated to a zoo remains "the same" tiger. This is the thinking that underlies the American saying that "you can take the girl out of the country, but you can't take the country out of the girl." From a Buddhist perspective, this idiom rightly points toward the mutual implication—the mutual enfolding—of persons and places. But according to its usual interpretation, it wrongly insists that the girl from the countryside who moves to the city always remains essentially a "country girl." This implies some degree of real independence. The Buddhist concept of interdependence goes substantially deeper: ultimately, there is nothing at all that exists independently, in and of itself.

To begin appreciating the radical sweep of this claim, consider trees. When I was a first-year graduate student preparing to introduce the idea of interdependence to local elementary students in a Philosophy for Children class, I asked my 7-year-old son the apparently simple question, "What is a tree?" His immediate response was to point out the living room window. Rather than defining what it is to be a tree, he offered me an example. Rephrasing, I asked him what it "took" to be a tree. Without hesitating, he described trees as having trunks and lots of branches and leaves, thus distinguishing trees from bushes and other plants. With a bit of prompting, he added that trees also had roots, like the one he'd tripped over at the playground the day before, and that to have trees you needed soil, sunshine, and rain.

To find out how imaginatively he was able to expand the horizons of what it takes to have trees, I then asked what would happen if you put a tree up in orbit—say, in the international space station. Would a tree remain a "tree" in zero gravity? Would it continue to have a canopy of skyward-lifted leaves supported by a more or less vertical trunk? Would its roots still grow "downward" if there were no longer any gravitational 'up' or 'down'? With these questions, he really got into the spirit of things. To have trees, you need the Earth. But the Earth wouldn't be the Earth without the Sun. With his library-fed imagination shifting into high gear, he described how the Earth orbits the Sun as part of one solar system among millions and millions of other solar systems in the Milky Way galaxy, which is itself just one galaxy among millions and millions of galaxies. To have a tree, he excitedly and confidently concluded, you needed the whole universe!

An "obvious" rejoinder is that while it's true that Earth's trees in some sense depend on the rest of the universe to exist as they do, the reverse surely can't be true. We can easily imagine trees (or, for that matter, human beings) ceasing to exist due to the destruction of the Earth by a massive meteor strike and this being just an infinitesimally minor event in a known universe that is ninety-three billion light years in diameter, with each light year equaling six trillion miles. But this imaginary scenario depends on taking up a kind of "view from nowhere" from which to observe the persistence of the cosmos without the Earth. If trees and humans were indeed eliminated, could it really be claimed that what remains is still the universe as we now know it? As sentient organisms, trees and humans are not just *objects in* the universe, they are *perspectives on* a world that only exists as such through them. To eliminate a perspective is to eliminate all that appears to exist objectively from it. Or, to put this somewhat more dramatically: any apparently objective universe exists only as a function or result of a specific point of view.

It is interesting to note here that something like this interdependence of the observed world and perspectives of observation is central to the physics of relativity and quantum mechanics that revolutionized modern science in the early twentieth century. As we will see in discussing karma in more detail, what Buddhism distinctively insists on is the sentient—that is, the feeling or affective—nature and quality of perspectival presence. Interdependence can be of different qualities, different "flavors." Thus, while realizing the interdependent or relational nature of all things is the essence of Buddhist wisdom, what makes that realization therapeutic is its marriage with compassion.

Emptiness (*śūnyatā*) and the Conceptual Nature of 'Things'

The statement that so-called objective reality is actually a function of perspective or point of view might be interpreted as an idealist claim about the ultimate reality of thought or experience: a denial of materiality. But idealist insistence on the independent existence of mind or spirit is not consistent with the insight that reality is ultimately relational. The Buddhist Middle Path, we should recall, runs athwart both idealist and materialist reductions. If the world prior to the assumption of perspectives *on* it can be characterized at all, it is as open or ambiguous. As the second century Indian Buddhist philosopher Nāgārjuna put it: the interdependence of all things means that they are ultimately empty (*śūnya*) of any kind of independent, essential, or abiding "self-nature" (*svabhāva*). To be unrelated is not to be at all.[3] Realizing the emptiness (*śūnyatā*) of all things means realizing their mutual relevance. To be is to mean something for others.

According to the main streams of Buddhist philosophy, we do not normally perceive the world this way because our experience is linguistically and conceptually conditioned. Words and concepts single out for attention different aspects of experience, attributing to them what amount to individual and at least relatively fixed identities. As I walk outside my front door and circle the house, I see an octopus tree, a mango tree, a lime tree, and an avocado tree. Above, there are some cumulus clouds. And across the valley is the extinct volcano, Mount Tantalus. Each of these appears to be an entirely separate entity. Walking is not the same as running; cooking is not the same as eating. But what these various words refer to are not independently existing entities or processes. They refer to different depths, qualities, and kinds of interaction—different patterns of dynamic relationality. The "Tantalus" of steeply pitched and thickly forested cinder cone that I clamber up has a weighty gravity totally absent in the "Tantalus" of winding roads and hidden residential cul-de-sacs that I explore in my pickup truck.

Reality, as it is constituted through the interactions of an ant with its surroundings, is not the same as reality constituted through the interactions of a human with the "same" surroundings. There are no 'clouds' as we know them in the ant's world, no 'mountains' or 'trees', and certainly no 'atoms' and 'molecules.' But the world of human interactivity does not include—except abstractly or through some technological proxy—the complex interplay of chemical signatures that are crucial to the ant's senses of location and presence. The cliff face that we can *see* is impossible to scale is, for the ant, a readily navigable landscape. Although it is a common belief that scientific and mathematical knowledge are getting better and better at revealing world "as it truly is," all that we are really justified in claiming is that scientific and mathematical descriptions bring into clearer and more detailed focus what human interactivity has been bringing into being *as* the world. The scope of our knowledge of the world may in some measurable sense exceed that of ants, but the human world is not ultimately any more real than the world of ant experience.

Concepts are distillations of stable patterns of interaction or relationality. Languages are media for interpersonal and intergenerational clarifications and elaborations of these patterns. That is, words do not refer to things in the world, but to *relational conventions*. Making distinctions between 'this' and 'that', or between what something 'is' and what it 'is-not', are not acts of *discovery*; they are acts of interest- or value-driven *disambiguation*. What each language makes evident—even the languages of mathematics and logic—is what Mahayana Buddhist philosophers have called "provisional" (*saṁvṛti*) reality, the world as we have enacted it, and not an "ultimate" (*paramārtha*) reality existing independently of our knowing relationships with it.

The differences among things are not intrinsic to *them*, but to *our interactions with them*. Or more accurately stated, the only boundaries that obtain among things are boundaries that we have imposed. Like the horizons we see as we turn slowly in a circle on a beach, these boundaries do not reveal absolute features of the world; they reveal features of our own perspectival presence. At every scale and in every domain—perceptual, cognitive, and emotional—everything we experience is the result of what we have elicited from our environments as actionable and as valuable. What *exists* for us is what *matters* for us. Indeed, matter is simply the definition of a point of view. Every existence marks the presence of a specific horizon of relevance.

To make this less abstract, consider the question of who comes first, parents or children? For most of us, the answer seems obvious. We might appreciate the logical tangle involved in trying to answer the structurally similar question about which comes first, the chicken or the egg. Nevertheless, our parents certainly preceded us and we certainly precede our own children. But in actuality, no one can exist *as* a parent prior to conceiving or adopting a child. Parenthood and childhood are coeval; neither can exist before or without the other. Moreover, what it means to be a mother, father, son, or daughter is not fixed. In Buddhist terms, parents and children are empty of any intrinsic and abiding self-nature. Our individual presences *as* fathers, mothers, sons, and daughters reveal what each of us identifies with as "me" and takes on as "my" roles in the dynamics of the family, and this will change dramatically over time. What matters and what it means to be a son at age eight or eighteen is not the same as at age 65 or 70. What "father," "mother," "son," and "daughter" refer to in any given family at any given time are value-expressing horizons of relevance, responsibility, and readiness.

Taken together, the Buddhist teachings of interdependence and emptiness do not offer a metaphysical claim about the nature of reality or "how things really are." Rather than instructing us as to *what* things are, they offer guidance in *how* we should see things in order to author our own liberation from conflict, trouble, and suffering. In other words, their purpose is exhortation, not revelation—encouraging us to depart from disputations about what 'is' or 'is not' real, true, mundane, or divine and to concern ourselves with the emancipatory significance of evaluating relational qualities.

Karma: The Meaning-Articulating Dynamics of Intentional Presence

By not seeing "reality" as something *discovered*, but rather as something *conferred*—a process of bringing or carrying together (Latin: *con+ferre*)—we effectively give precedence to ethics over both epistemology and metaphysics.[4] Who we *should* be present *as* has precedence over who we presently *are*, or who we have been, because our personal realities are the dynamic results of our most consistently held values and embodied intentions. Our realities are karmic.

Given the Buddhist understandings of experience, interdependence, and emptiness, it follows that conflict, trouble, and suffering cannot be held to be matters of chance or fate in which we are only contingently implicated. Conflict, trouble, and suffering result from the ways in which we have disambiguated things—the ways in which relational dynamics have been inflected by us acting on the basis of where we have set the boundaries between what is relevant and what is irrelevant, between what we accept responsibility for and what we do not, between what we are ready to engage in and what we are not, and between what we desire or feel we need and what we do not. In traditional Buddhist terms, *duḥkha* is a result of enacting our own ignorance, habits, and cravings. In whatever ways we conventionally experience and understand them, conflict, trouble, and suffering are ultimately markers (*lakkhaṇa*; *lakṣaṇa*) of relational disruptions and distortions that are caused and conditioned by conceptual proliferation (*papañca*; *prapañca*) and by the corporeal, communicative, and cognitive conduct we engage in on the basis of it. *Duḥkha* is an expression of our own karma.

Karma has been commonly understood as a synonym for destiny or fate. But as it was originally formulated in Buddhist teachings, karma is not about fate; it is about dramatic continuity. In accounts of the night during which the Buddha realized enlightenment and liberation from *duḥkha*, he is depicted first as attaining two forms of meditation-generated "super-knowledge" related to karma: a direct perception of his present life as part of a dramatic genealogy stretching back over numberless lifetimes, and a more general perception of how the birth circumstances and experiential dynamics of all sentient beings are inflected by their own patterns of intention and conduct. While these accounts are silent with respect to *how* this karmic conditioning occurs, *that* it occurs is crucial to their affirmation of the compassionate aims and emancipatory potential of Buddhist practice.[5]

While the Sanskrit word *karma* (Pali: *kamma*) has the literal meaning of action, it derives from the same Indo-European root as the English word "drama." In keeping with this root, karma does not refer, for example, to the mechanical action of wind, rain, and waves eroding an oceanfront cliff. Karma implies agent-originated action: opening a wide channel through a near-shore reef to create a boat harbor or building ocean-view homes along the edge of a sandstone bluff.

In pre-Buddhist Indian religious thought, "karma" came to designate a cosmic process or moral law that determines the future contexts of one's exercise of agency. More specifically, one's actions were believed to determine whether one's future births would be as a hell-denizen, a hungry ghost, an animal, a human, a titan, or a god. This causal process of karmic determination was understood to be simple and linear: violent acts necessarily beget future experiences of violence, while kind acts beget future experiences of kindness. Karma was the operation of a law of just exchange in a cosmic moral economy.

The Buddhist innovation was to emphasize the role of volition and to delink karma from any transcendent moral order. Karma, in Buddhist terms, is the ongoing articulation, from within, of *emergent moral orders* in which intentionality (*cetanā*) and values play crucial roles. Rather than a transcendent or objective system for guaranteeing our delivery into fates we have crafted through our own past actions, the Buddha saw karma as a reflexive or recursive process through which we articulate our own experiential possibilities. In other words, rather than delivering us into the destinies or fates that we deserve, the operation of karma is precisely what warrants confidence in our capacities for changing the dramatic tenor and tendencies of experience.

In combination, the Buddhist teachings of interdependence and karma enjoin us to see that it is as impossible to establish the priority of mind or matter as it is to establish which came first, the fruit or the tree. Mango fruits and mango trees are equally basic aspects of an ongoing pattern of relational dynamics that comprises both biological and environmental dimensions. Mind and matter are similarly aspects of a seamlessly ongoing process of karmic becoming. Sentient beings (actors/observers) and their environments (things acted-on/observed) are like the two sides of a Möbius strip, a three-dimensional figure that has only one side and one edge. Ultimately, there is no gap between "matter" (the space of causes) and "what matters" (the space of intentions). All action is interaction.[6]

By paying close and sufficiently sustained attention to the dynamics of our own life experiences, it becomes evident that the experienced outcomes of our actions—mental, verbal, and physical—and the opportunities they bring are unfailingly correlated with consistencies in the values and intentions informing our actions. Through doing so, as the Buddha is said to have done, it becomes clear that the world we inhabit is irreducibly meaning-laden, but also that it is not a world of scripted interactions. Our experienced realities are improvised dramas, playing out live—dramas for which we are responsible together and in which we thus always have real opportunities for changing the way things are changing.

Our natural tendency is to think of these as opportunities to change the objective conditions causing us to have unwanted experiences. The Buddhist teaching of karma invites us to think otherwise. It is safe to assume that it is never our direct intention to bring about conditions that will cause us to experience conflict, trouble, and suffering. Given this, and given that our experience is indeed karmically ordered, the fact that we *do* experience conflict, trouble, and suffering has to be understood as evidence of tensions between enacted values and intentions that individually would bring about desirable consequences, but that in combination generate troubling relational and experiential cross talk. In short, our experiences of *duḥkha* are evidence of the predicament-laden nature of our experience. Changing the *way* things are changing is ultimately not a matter of altering this or that aspect of our objective circumstances. It involves the much harder labor of predicament resolution.

To anticipate why the concept of karma might be particularly relevant for clarifying the risks of a craving- and control-driven attention economy, as well as the more general value conflicts accentuated by the Intelligence Revolution, consider how our internet experience is now shaped by self-improving algorithms (decision-making procedures) that use patterns of our searches, preferences, and purchases to determine and structure the web content made available to us. These algorithms are designed to continuously shape and reshape our online experiences to bring them into ever closer accord with what we would like, based on our own expressed aims and interests. In short, the values embodied in our connective behavior are used recursively to configure and progressively reconfigure our online realities. In effect, the computational factories of the Intelligence Revolution are functioning as karmic engines.

In theory, the web affords us access to everything digital. In actuality, algorithmic filtering is crafting the web content we experience so that it becomes ever more suitably and uniquely our own. The values we express and act upon set the horizons and content of our future experience. To play with a phrasing gleaned from the Mahāyāna Buddhist text, the Diamond Sutra: our 'everything' is not everything, even though we have no reason not to refer to it as "everything." It *is* all that we are being enabled to experience, even though it does *not* include everything that could be referred to *as* "everything." Algorithms are functioning as arbiters of our "digital karma."

The teaching of karma is that our offline lives, just like our online lives, are shaped by patterns among our values and intentions. The values that inform our thinking, speaking, and acting establish our life headings, specifying or bringing into actionable focus only some aspects of all that is (or could have been) present. But our computationally engineered karma—the shaping of experience that is being undertaken by ambient machine learning systems—is not geared to present desired outcomes in ways that also afford opportunities for changing our values, intentions, and actions in response to any troubling patterns of interdependence in which they might be implicating us. On the contrary, these surrogate machine agents are geared toward figuring out how to give us only and always what we want.

The ethical concern is not just that we are being enabled to live in "filter bubbles" (Pariser 2012), insulated from having to confront others with different values and interests. Conflict avoidance is an all-too-common tactic for managing our experience and requires no digital assistance whatsoever. The concern is that decisions about whether to avoid others with different views are being made for us. If they continue developing as they are at present, the intelligence industries and their computational factories have the potential to become so sure and seamless in anticipating and providing us with what we like and desire that we will lose the opportunity to make and learn from our own mistakes.

The technologically achieved impossibility of making mistakes and being disappointed in choices we have made might initially seem appealing. But reflecting for a moment on what it would mean to become effectively "locked in" to the patterns of values and interests we had as teenagers or toddlers should suffice to see the very real dangers involved. What is at stake, ultimately, is the opportunity to engage in the intelligent human practice of predicament resolution, developing the creativity and moral maturity needed to participate effectively in intercultural and intergenerational ethical improvisation.

Buddhist Practice: The Teaching of the Three Marks

The purpose of Buddhist practice is to change who we are present as in order to be effective in realizing more liberating patterns of change and freeing ourselves from *duḥkha*. The so-called teaching of the three marks (*tilakkhaṇa*; *trilakṣaṇa*) is one of the most succinct formulations of how to do so. Although it is often represented as a set of metaphysical doctrines about how things *are*, the teaching of the three marks is more aptly understood as a set of practical directives for dissolving the conditions for conflict, trouble, and suffering: seeing all things *as* troubled or troubling (*duḥkha*); *as* impermanent (*anicca*; *anitya*); and *as* without essence/self (*anattā*; *anatman*). It is, in other words, a method for revising *how* we are present: a method for relinquishing the fixed standpoints from which we have habitually demarcated 'me' and 'mine' from 'you' and 'yours', qualitatively transforming our dynamic interdependence from the inside out.

Seeing All Things as Troubling. The teaching of the three marks suggests that seeing all things as marked by conflict, trouble, and suffering is a method for eliminating the conditions that give rise to these undesirable experiences. But for most of us, accustomed as we are to evaluating things on the basis of our individual perspectives, the invitation to see all things as marked by *duḥkha* does not make intuitive sense as a way to eliminate the conditions that give rise to it. In fact, it can seem like an invitation to adopt the relentlessly pessimistic view that "life is suffering." This interpretation, however, is consistent neither with traditional presentations of the teaching of the three marks nor with the aim of Buddhist practice.

To begin with, it is undeniable that the flavors of a hearty American breakfast of bacon, eggs, and toast are pleasurable. Falling in love is a glorious feeling. A walk along the seashore, listening to waves washing up and then percolating down into the sand is deeply relaxing. Empirically, it is simply not the case that "life is suffering." The point of seeing all things *as* troubled/troubling is not to foster abject pessimism. It is to practice shifting attention according to context and across scales of interdependence, verifying the falsity of assuming that if "I'm okay, you must be okay." It is, in short, a practice of cultivating ethical awareness.

To begin appreciating how this works, consider the mundane example of having bacon and eggs for breakfast. Only a slight shift of perspective is needed to see that this breakfast means something different for humans, hogs, and chickens. Our pleasure while eating bacon comes at the cost of the hog's life. While chickens do not have to die for us to eat their eggs, if they are not free-range chickens, their industrially constrained lives are spent more or less immobilized as biological "machines" in "factories" engineered to produce eggs at the highest possible rates and for the lowest possible unit cost. Industrial agribusiness treats animals not as ends in themselves but as mere means to meeting human needs for nutrition and pleasure, doing so in ways that satisfy corporate desires for profit.

Even if we dismiss the rights of animals to moral consideration, however, our breakfast of industrially produced bacon and eggs can still be seen as troubling. The more successfully the goals of industrial agribusiness are met, the more difficult it is for family farms to remain viable as businesses without adopting similar cost-cutting practices. Moreover, when eating bacon and eggs is promoted as a nutritional ideal for the purpose of increasing agribusiness profits, the resulting negative health impacts can be quite pronounced. Eating bacon and eggs in moderation can be both pleasurable and nutritious. But being indoctrinated by corporate advertising to eat bacon and eggs every day can easily have health-compromising, if not life-threatening, impacts. Seeing a bacon and egg breakfast as *duḥkha* is a method for bringing ethical considerations to bear on the pleasures of eating and the structural dynamics of meeting daily nutritional needs.

The point of seeing all things as implicated in patterns of interdependence characterized by conflict, trouble, and suffering is not to stop enjoying life. The point is to expand the horizons of responsibility, within the compass of which we make life decisions, determining what to do and what to refrain from doing. Our actions are never simply ways of *effecting* our own aims and interests. Our actions are always also *affecting* others and the world around us in ways that will eventually affect us in return. This experiential feedback can occur because others interpret and directly respond to our actions. Or it can occur through systemic transformations like those brought about by environmental degradation and climate change. Either way, acting on others and the world around us is acting on ourselves. There are no agents that are not also patients of their own actions.

This means that at least when other sentient agents are involved, there is always an emotional aspect to the feedback loop joining intentional conduct and subsequent experiential outcomes and opportunities. Our behavior is *felt* and responded to by others as wanted or as unwanted, as pleasant or as unpleasant, as consonant or as dissonant with their own nature and interests. In fact, even an inanimate piece of wood will "resist" our artisanal efforts if care is not taken to align our interests and actions with the wood's grain—the record of its formation and deformation over time. We can, of course, use power tools in carving wood. We can override others' interests by employing physical, social, legal, or other powers in excess of their capacities for resistance. We can force things to go our way. But as we will see very clearly in discussing the human-technology-world relationship, the karmic result will be experiences of new and deeper kinds of resistance and resentment, and thus multiplying opportunities, if not compulsions, to exercise still greater power.

Alternatively, we can refrain from exerting power over others. Doing so will enable us in many instances to enjoy experiences of resonant mutuality—a sharing of aims and interests accompanied by feelings of accelerated movement in desirable directions. But this clearly is not always the case. Consider, for example, getting a workplace promotion, or winning a prestigious scholarship, or simply attracting someone's romantic interest. In each case, even if our happiness was not actively sought at others' expense, there will often be others who will be deeply disappointed when things go decisively our way and not theirs. When one company gains overwhelming market dominance, others are at risk of failing to remain solvent; when one country achieves overwhelming military might, others fear aggression and act accordingly. To see all things *as* troubled or troubling is not to indulge in abject pessimism; it is to see all experience in terms of mutually affecting, meaningfully felt, and hence ethically charged relationships.

Seeing All Things as Impermanent. Seeing all things as impermanent undermines expectations that good situations will last forever. The pleasing sense of fullness after a fine meal passes away into new hunger. The sensual blossoming of romantic love does not last a lifetime. Health is interrupted by illness. Getting your dream job is not the same as doing that job day in and day out, month after month, year after year. At some point, to quote the old blues song, we discover that "the thrill is gone."

Seeing all things as impermanent is, at a deeper level, to practice seeing all things as processes—as emergent phenomena within always ongoing relational dynamics, rather than as stably existing entities that are either shoved about by the winds or currents of change or as abiding agents that somehow manage to ride them. The Greek philosopher Heraclitus famously proclaimed the impossibility of stepping into "the same" river twice since the waters comprising it are always flowing onward. To this, Buddhism adds that we as observers and the bridges or rocky promontories from which we view the river are also always "flowing on." Bridges and rocky promontories may "flow" more slowly than humans, but they age and change character no less certainly than we do over time. What sets humans and other sentient beings apart are our capacities for affecting the pace, direction, and qualitative dynamics of change. We are not merely beings who are *subject to* change; we are ever-becoming *subjects of* change.

Most importantly for Buddhist practice, seeing all things as impermanent also entails recognizing that no bad experiences last forever and that no situations are truly intractable. If change is always ongoing, there can never be any real question about whether change is possible. What is uncertain are only the direction, pace, and quality of change. Seeing all things as impermanent is thus training to be aware that it is always possible to change *how* things are changing. Even when we find ourselves apparently stuck in a troubling pattern of interaction, the *meaning* of this pattern—the experiential consequences of its apparent persistence, perhaps in spite of our sincere efforts to break free of it—is never fixed. The significance of our situation is always negotiable.

It follows from this that whenever we claim there is no way for us to change some situation, we are in fact proclaiming that we are only subject to—and not also the responsible and responsive subjects of—our experience. Seeing all things as impermanent involves seeing how intractability never belongs to our circumstances, but to the fixed nature of the positions we assume within them. The experience of being stuck where and as we are announces our failure or refusal to attend to our situation as one in which we are openly and dynamically implicated. Karmically, the feeling that there is nothing we can do to change our situation is self-justifying evidence of our own resistance to changing who we are present as.

This is perhaps easiest to see in others. The typically American battle of wills fought between teenagers and their parents, with one side professing individual rights and the other familial responsibilities, is a struggle over how to change the course and quality of familial dynamics. This is a battle that will either be won by both or lost by both. Seen from a friendly distance, it is usually clear that the central determinant of how things will turn out is not *what* changes the parties involved are individually calling for, but *how* they are doing so.

In one of the earliest written Buddhist texts, the Sutta Nipāta, the Buddha identifies the core conditions for persistent conflict, trouble, and suffering as the belief that "this is true, all else is false" and the conviction that conflict-settling naturally results in distinct winners and losers (see, e.g., Sutta Nipāta, IV.8, IV.11 and IV.13). At the root of our incapacity to change how things are changing to alleviate conflict, trouble, and suffering is the certainty which results from adhering to a fixed perspective on the world, combined with a belief in the ultimate reality—or at least the ideal—of our independence as essentially autonomous individuals. In short, the root cause of conflict, trouble, and suffering is attachment to a fixed identity or sense of self.

Seeing All Things as Without-Self. In a karmically ordered cosmos, seeing all things as without-self or as lacking any abiding essence is not an exercise in nihilism. It is a practice of regarding the meaning of all things, including our own selves, as open to revision. To see all things as empty (*śūnya*) of any essence or fixed nature is to realize that none are intrinsically either good or bad, and that no situation is bereft of resources for bringing about more enlightening or liberating patterns and qualities of relationality.

Seeing "other" things as without-self is not that difficult. In the practice of seeing things as constantly changing, we are very close to seeing them as lacking any essentially fixed identities. We quickly discover that rather than being a kind of *warehouse* of individual entities persisting over time, the world is a *composition* of processual continuities. Although we can say that "the" Colorado River has "existed" for fifteen or twenty million years, this is just a convenient way of speaking. The river's course has changed dramatically over that period of time and for the last five million years has been cutting an ever deeper channel—the Grand Canyon—into the Colorado Plateau. Fed by rain and snowmelt gathered from nearly a quarter million square miles of mountainous terrain that is now tilted toward the Gulf of California, the Colorado River is an ongoing composition—a cyclical, hydrological singing—of planetary climatic, tectonic, and biological forces. What we refer to as the Colorado River is without-self.

But as soon as we try to extend the practice of seeing all things as without-self to ourselves, matters are neither so self-evident nor so simple. We can readily accept that the sprouting mango, the sapling, and the mature tree are not "the same" in either substance or structure. We can see the mango tree before us as part of an evolutionary process in which not even the mango genetic code is permanent—a process that we can literally manipulate through cross-breeding. But when it comes to our own presences, we are much less ready to give up on the existence of an abiding self.

We might be willing to admit that every cell in our bodies is replaced roughly every seven years. Our bodies are not the same as they were a decade ago. But isn't this still 'my' body and that one 'yours'? In the end, doesn't there have to be a permanent 'seer' who is continuously present to see our bodies and other things as impermanent? Doesn't there need to be an abiding thinker or 'me' that links passing thoughts together as 'mine'? Don't my memories have to adhere to some continuing locus of experience in order to be experienced as 'mine'? Your normal waking experience, like mine, is that of being a unique and abiding center of awareness, emotion, judgment, and action. Each of us has a name, characteristic likes and dislikes, and a particular place in a unique family and community. We are not interchangeable. Our experiences are not identical. Given all this, what could possibly be meant by saying that we are without-self?

As a way of answering such questions, the Buddha simply invited people to try locating an abiding self in their own experience. As anyone who tries doing so quickly and perhaps somewhat surprisingly discovers, no such self is anywhere in evidence. Instead of a central and abiding self or soul, what we find are five clusters of experiential/relational dynamics (*khandha*; *skandha*). These clusters center on being bodily present (*rūpa*); on being present as feelings of liking, disliking, and neutrality (*vedanā*); on being present perceptually through seeing, hearing, smelling, tasting, touching, and thinking (*saññā*; *saṃjña*); on being present volitionally, actively, as well as habitually (*saṅkhāra*; *saṃskāra*); and on being present as one of the six modes of discriminating consciousness that emerge through the interactions of the six sense organs and objects sensible to them (*viññāṇa*; *vijñāna*).[7]

No matter how diligently we investigate each of these clusters of experiential presence, we find nothing that accords with the concept of self as an unchanging, underlying substance or entity. All that is evident is an ever-shifting composition of bodily, emotional, perceptual, volitional, and discriminatory occurrences—a play of interdependent phenomenal events. These events may all seem to be connected to 'me' in some way, but this 'me' is itself nowhere to be found. Like a cluster of asteroids spinning around one another in interplanetary space, the five *skandhas* of sentient presence are joined in something like an orbital dance around a "missing" center of attentive gravity. The self—the presumed 'owner' of experience—is only virtually (and not actually) present.

This Buddhist practice of seeing our personal presence as a composition of five ever-shifting clusters of mutually dependent experiential factors has been compared to the so-called bundle theory of self that was forwarded by David Hume (1711–1776) in his *Treatise on Human Nature*, more recent versions of which have been articulated by thinkers like Daniel Dennett, Derek Parfit, and Owen Flanagan. But whereas these so-called reductionist theories of self effectively deny the presence of a lasting moral agent, equating being without-self with being without-responsibility, the Buddhist concept of no-self or being without-self is intimately allied with teachings about the centrality of karma and compassion in the therapeutic system of Buddhist practice.[8] In fact, the Buddha was often confronted by critics who had interpreted the practice of seeing all things as without-self in reductionist terms and who then pointedly challenged him to explain how karma could operate in the absence of an agent that persists over time and is the moral patient of its own past actions. It's instructive that in these encounters, the Buddha resolutely insisted on according primacy to therapy (what works) rather than to theory (what can be explained) and often remained silent as a way of letting critics know that their questions were unanswerable in the terms that they were posed.

Among these questions were: "Do I really exist now? Did I exist in the past, and if so who was I and how did I live? Will I live again in future lives and who will I be? Where did I come from and where am I ultimately bound?" These questions are unanswerable because responding to them involves committing to some *duḥkha*-generating point of view. "I have a self. I have no self. It is precisely by means of self that I perceive self. It is precisely by means of self that I perceive not-self. It is precisely by means of not-self that I perceive self. This very self of mine—the knower who is sensitive here and there to the ripening of good and bad actions—is constant, everlasting, eternal, not subject to change, and will stay just as it is for eternity" (*Sabbasava Sutta*, MN 2). However different these points of view may be theoretically, they are alike in being therapeutically counterproductive.[9] Adopting any of these positions only results in being caught up in birth, aging, and death; in pain, distress, and despair; and in conflict, trouble, and suffering.

Understanding the teaching of no-self in its original soteriological context is important insurance against concluding that being without-self prohibits having purposes, values, or bases for making and keeping commitments. Being without-self is not *being absent*. It is training to be present so that "in the seen there will only be the seen; in the heard, only the heard; in the sensed, only the sensed; and in the cognized, only the cognized." Being present in this way, "no 'you' is *with that*; when no 'you' is *with that*, no 'you' is *in that*; and when there is no 'you' either *with that* or *in that*, there is no 'you' here, there or in between the two, and it is precisely this that is the end of *duḥkha*" (Bāhiya Sūtta, *Udana* 1.10). Being without-self is *not* being present *as subject to* suffering, conflict, and trouble. It is not the absence of continuous intentional moral agency, but only that of the abiding moral agent who might be permanently damaged by conflict or scarred by experience. The practice of being without-self is one of becoming present as needed for the emergence of liberating agency.

Some sense of what this practice involves can be gleaned from reflecting, for example, on the moment one first learned to ride a bicycle. I learned from my older brother, who had me sit on the bike and begin pedaling as he steadied the handlebar, walking and then jogging next to me for a few yards before releasing me to continue on my own. The unfailing result was a rapid loss of balance and a crash onto the concrete sidewalk or the grass beside it. But then everything clicked. My brother let go of the handlebar and suddenly there was no thinking about peddling and keeping the handlebars steady and staying on the sidewalk, no worrying about falling. "I" vanished and all that was left was balance in movement, peddling freely and easily, faster and faster. At moments like this, we are neither in control nor out of control. We are present only as fully embodied and wholly active (in this case, bike riding) agency.

This achievement of total immersion in an activity can occur in almost any kind of activity—in doing art, in sports, in musical performance, in manual or mental labor. It is generally the result of intense effort applied over time in a consistent practice, resulting in what the psychologist Mihaly Csikszentmihalyi (1990) has termed optimal experience or "flow"—an exhilarating realization of activity so deeply and thoroughly concentrated as to bring about one's complete absorption in it. Usually, however, this is not a transferable achievement. "Flow" is a domain-specific, challenge-handling, and striving-generated achievement. A virtuosic musician or world-class athlete may be able to somewhat regularly disappear into the "flow" of peak performance and yet be wholly incapable of doing so outside the concert hall or off the playing field. In other parts of their lives, they may continue to find themselves blocked or caught by their circumstances, acting neither wisely nor compassionately. The purpose of Buddhist practice is to be present without-self in all circumstances, realizing ever-deepening accord with one's circumstances, and responding as needed to bring about more liberating relational dynamics, benefitting all involved.

An Ethics of Compassionate Relational Virtuosity

If ethics is the art of human course correction, Buddhist ethics is course correction based on the integral embodiment of wisdom (Pali: *paññā*; Skt.: *prajñā*), moral clarity/discipline (Pali: *sīla*; Skt.: *śīla*), and attentive mastery (Pali and Skt.: *samādhi*). It is their joint cultivation that enables skillfully and successfully directing relational dynamics away from the affective distortions of conflict, trouble, and suffering, and toward nirvana (Pali: *nibbāna*), the consummate aim of the so-called Eightfold Path of Buddhist practice: engaging in correct/corrective action, speech, and livelihood as part of cultivating moral clarity (*śīla*); correct/corrective effort, mindfulness/remembrance, and attentive poise as aspects of cultivating attentive mastery (*samādhi*); and correct/corrective views and intentions as dimensions of cultivating wisdom (*prajñā*).

Nirvana literally means "blown out" or "cooled down." The first modern European and American interpreters of Buddhism, grappling with the foreignness of teachings that stressed being without-self and realizing the emptiness of all things, can perhaps be forgiven for concluding that Buddhism was a nihilistic religion in which salvation amounted to being snuffed out like a candle flame. In fact, the extinguishing of a candle flame was a common metaphor used by many early Buddhists to explain nirvana. But the point of the metaphor was to direct attention to the *process* by means of which the "flames" of conflict, trouble, and suffering are put out. Just as removing a burning candle's wick or the oxygen surrounding it will result in its flame disappearing, the "flames" of *duḥkha* will be extinguished by removing the conditions of belief in independent existences and captivation by clinging forms of desire. Thus, rather than the goal of Buddhist practice, a destination to be arrived at, nirvana is its therapeutic orientation.

It is one of the distinguishing features of Buddhist ethics that its ultimate aim, the "good" toward which Buddhist practice is oriented, is not positively characterized and remains resolutely undefined. Buddhist ethics offers no conceptual maps or fixed principles for arriving rationally at the "good life" or building a "good society." Instead, consistent with the practices of seeing all things as implicated in conflict, trouble, and suffering, as changing and as lacking any fixed identities, implicit to Buddhist ethics is an acceptance of the fact that the course corrections that may be warranted in any particular situation could not have been determined in advance. Indeed, the very possibility of attaining nirvana is predicated on the karmic fact that the way things are changing is conditional and always open to change. Buddhist ethics consists in *skillful* course correction in the absence of a "moral telescope" that might allow us to see in advance where we should be going.

This skill depends on the insights and wisdom that emerge through the continuous and deepening practices of seeing all things as interdependent, as karmically configured, and as without-self. Buddhist wisdom is not something achieved *through* acquiring specific bodies of knowledge or through enduring the perspective-widening processes of aging and maturation. It is an achievement *of* steadfastly relinquishing the horizons of relevance, responsibility, and readiness that until now have *defined* who we *are* and thus limited who we have been capable of being *present as*.[10] Buddhist ethics involves *furthering* that process. In carrying out relational course corrections, wisdom does not function as a "moral telescope," but as a "moral compass."

Successful course correction also requires a very clear understanding of current conditions. It is one thing to know that every situation we find ourselves in is an expression of some karmically configured pattern of interdependence and that relational turbulence is ultimately the result of conflicting values and intentions. It is another to discern and correctly read the currents of intention and value that are implicated in *this* situation and exactly *how* they are affecting relational dynamics. It is not possible for a sailor to keep on course without being keenly sensitive to even the most subtle shifts in currents and winds. That is not possible if he or she is daydreaming or drunk. Cultivating and maintaining keen *sensitivity* to karmic currents and the winds of passions and desires is crucial to Buddhist ethics.

As part of the moral discipline involved in cultivating and maintaining moral sensitivity, all Buddhist practitioners take five vows: to refrain from harming or killing others, from speaking in hurtful ways, from sexual impropriety, from using intoxicants to the point of heedlessness, and from taking what was not freely given. In much the same way that basic hygiene practices like regularly washing our hands and cleaning our homes can prevent us from catching and spreading many common contagious diseases, refraining from these actions works as a kind of karmic hygiene that ensures basic "moral health." But maintaining a clean body and home is no guarantee of optimal health. In addition, an exercise regime and supportive diet may be needed as well. Keeping the five precepts is good, but realizing liberating relational dynamics will require also adequately reading and responding to the karmic currents implicated, for example, in emotional, cognitive, social, cultural, or political conflicts and turbulence: the exercise of attentive capacities that are almost athletic in their focus and flexibility.

To embody wisdom and enact moral clarity requires attentive mastery. We will later discuss the roles played by focus- and flexibility-oriented meditation practices in realizing Buddhist ideals of personal presence. Here, anticipating critical engagement with the dynamics of the attention economy, it is enough to stress that attention training is integral to the processes of physical, emotional, and intellectual dehabituation that are needed to be freely responsive. The Pali and Sanskrit term for attention, *manasikāra*, simply means awareness that is concentrated or resolutely focused. This implies that one can be attentive with different degrees of concentration or focus. We can devote half our attention to cooking and half to conversing. But in addition to how much attention we are paying to our situation, Buddhism makes a distinction qualitatively between being attentive in ways that bind us to or that free us from conflict, trouble, and suffering.

It is possible, even without training, to be keenly attentive to our present circumstances. Young children avidly awaiting the ice-cream cone being prepared for them and adolescents in the throes of video game ecstasy are both clearly capable of highly concentrated attention. What is not so clear is whether they are freely attentive or compulsively so. Without training, our attention is readily and *involuntarily* attracted or distracted. In particular, we are especially susceptible to unwisely having our attention captured by the superficial, craving-inducing aspects of things (*ayoniśomanasikāra*). This, as we will see, is crucial to the workings of the new attention economy being realized through intelligent technology. Yet, with training, our attention can also be wisely concentrated—directed freely and *intentionally* in ways that are both sensitive to the interdependent origins of things and consistent with *truing* relational patterns (*yoniśomanasikāra*).

To the extent that Buddhist ethics consists in the goalless, nirvana-oriented practice of integrally cultivating wisdom, moral clarity, and attentive mastery, it is hard to place readily or without remainder into one of the standard categories of ethics grounded on definitive and generalized judgments regarding personal character (virtue ethics), duties (deontological ethics), or the consequences of actions (utilitarianism). Given Buddhism's ethical insistence on pairing wisdom with compassion, a closer fit might be care ethics, with its emphasis on situationally apt attentive responsiveness. But Buddhist compassion is not reducible to the natural inclinations to care about and for others that are invoked by care ethics, much less to the abstractly mandated responses to suffering that are typically framed with reference to personal virtues or duties, or derived through a consequentialist calculus of harms and happiness. Rather, Buddhist compassion is exemplified in the ongoing intentional practice of dissolving the karmic causes and conditions of shared conflict, trouble, and suffering—a necessarily improvisational labor of shared predicament resolution in steadfast pursuit of increasingly liberating relational outcomes and opportunities.

What makes Buddhist ethics so difficult to place (and, potentially, so relevant today) is the fact that it offers only an open-ended training program—cultivating wisdom, moral clarity, and attentive mastery—and a set of "cardinal points" for discriminating qualitatively among relational outcomes and opportunities. Especially in early Buddhist contexts, the term used for the "true north" of liberating presence on the Buddhist "moral compass" was *kuśala*. Often translated as skillful or wholesome or good, *kuśala* actually functions as a superlative. Rather than connoting something that is good as opposed to mediocre or bad, it connotes virtuosity.

The ethical significance of aiming at *kuśala* outcomes and opportunities is neatly illustrated in an early Buddhist text, the *Sakkapañha Sutta* (DN 21). Like most early Buddhist suttas or recounted teachings of the Buddha, the *Sakkapañha Sutta* is structured as a dialogue. In this case, the Buddha is asked to explain how it can be that human beings generally want to live in harmony and without strife, and seem to have the resources for doing so, they almost always fail and end up embroiled in anger, hatred, and conflict. At first, the Buddha offers his standard psychological account of conflict and social strife as typically being rooted in jealousy and greed, which are in turn dependent on having fixed likes and dislikes, and these on being caught by craving forms of desires and tendencies to dwell on things. But this entire edifice of conditions, he finally explains, ultimately rests on conceptual proliferation (Pali: *papañca*; Skt: *prapañca*): compulsively dividing up what is present into ever more finely wrought units and relations among them, producing ever more tightly woven nets of fixed associations and judgments that at once support and entrap the craving- and conflict-defined self. To bring an end to conflict, interpersonal discord, and the suffering they entail, one must uproot *prapañca*.

When the Buddha is asked how we can stop engaging in conceptual proliferation and enact our intentions to live in peace and harmony, he significantly directs attention away from "inner" psychological conditions to "outer" personal and social consequences. To cut through *prapañca*, he says, we should continually evaluate our conduct (mental, verbal, and physical) in terms of whether it is bringing about *kuśala* or *akuśala* outcomes and opportunities, continuing on courses of actions only if they both decrease *akuśala* eventualities and increase those that are *kuśala*. Given that *kuśala* is a superlative, this means that resolving conflicts and freeing ourselves from trouble and suffering is not simply a matter of refraining from doing bad things and instead doing or being either harmlessly mediocre or what is considered good by current standards. These are all *akuśala*. Freeing ourselves from conflict, trouble, and suffering requires going beyond current conceptions of good and evil, realizing virtuosically shared presence with and for others. The course correction required is resolutely qualitative.

The aim of Buddhist ethics is to foster the cultivation of wisdom, moral clarity, and attentive mastery, establishing and then continuously enhancing commitments to and capacities for thinking, speaking, and acting as needed to realize superlative or virtuosic (*kuśala*) relational dynamics. The purpose of ethical deliberation is not to discover or devise absolute or universal standards of conduct. Just as virtuosic musical performances set new standards of musicianship, *kuśala* ethical conduct sets ever new standards of ethical excellence. A karmic ethics of compassionate virtuosity is an ethics of doing better at what we are already doing best, evaluating value systems and the ways that they are embodied personally and institutionally to realize ways of life that are progressively conducive to relating freely.

*

The concise introduction to Buddhism just offered cannot do justice to the philosophical or religious complexity of the Buddhist "ecologies" that have been evolving for over twenty-five hundred years. Its purpose has been to foreground a set of sensitivities that will be helpful for freshly engaging the history of intelligent technology over the next chapters as a history, not only of brilliant inventors and their inventions but also of complex aspirations and societal drives. These include sensitivities to interdependencies, for example, among scientific, commercial, and political agendas; to the ways in which consistently enacted values and intentions bring about equally consistent patterns of outcomes and opportunities, accelerating feedback and feedforward loops that can lock in biases as well as open new horizons; and to the dynamically complex interfusions of agency and structure that shape the human-technology-world relationship. What would it mean for that relationship to be superlative as technology becomes increasingly intelligent?

It is now nearly a standard practice in mainstream ethics of AI and related technological advances to stress the importance of design, development, and deployment that are human-centered. There is an undeniable appeal and logic to this. Although future biotechnology and AI may allow otherwise, we cannot now be other than human. We may make provocative claims about our transhuman futures (Manzocco 2019) or about according rights to robots (Gunkel 2019) and other abiotic and biotic entities (Harraway 2015), but we can only do so from our perspectives *as* human. Buddhism adds to this commonsense admission of the primacy of the human in ethical considerations an invitation to begin seeing that we also cannot be other than interdependent. Granted this, in evaluating the interdependencies that are constitutive of the human-technology-world relationship, Buddhism invites looking beyond binary considerations of whether or not intelligent technology is human-centered to consider how best to ensure that humanity and intelligent technology are as *humanely* interdependent as possible.

Prior to the eighteenth century, the word "humane" was simply a variant for "human." It was only in the eighteenth century, roughly at the beginning of the First Industrial Revolution, that "humane" came to signify kindness, compassion, and benevolence—qualities that were for the first time seen as attributable, not only to people but to actions, processes, and institutions. Given the open-ended nature of Buddhist ethics, these qualities should not be considered either definitive or exhaustive of a superlative human-technology-world relationship. But at least provisionally, they afford a footing on which to frame explicitly qualitative (if not fully normative) concerns about intelligent technology. As the following chapters will make evident, the Fourth Industrial Revolution—an Intelligence Revolution—is well underway and rapidly accelerating, almost miraculously scaling up human values and intentions. Given the stakes involved, it is fortunate that whether it will prove to be a truly humane revolution is still to be determined.

2

Artificial Intelligence: A Brief History

When we think of the conflicts of interests arising with the emergence of intelligent technology, we think of the future. Can machine intelligences be developed in ways that align with human values? Will artificial intelligence (AI) surpass human intelligence one day? How will the Intelligence Revolution change the lives of our children and grandchildren? Will the internet of things make us more secure or more vulnerable? These are all forward-looking questions. The premise of this chapter is that if we want to be able to influence what the Intelligence Revolution will mean for the human experience, we need also to look backward.

One of the core insights of the Buddha was that we can only effectively and sustainably resolve the conflicts, troubles, or suffering that we are experiencing on the basis of first understanding how things have come to be as they are. A "snapshot" of the present is not enough, no matter how wide-angled the lens or how detailed the image. We need a "film," and ideally one with "footage" shot from many different angles. Histories matter.

Servants of Our Own Making: Dreams of Artificial Beings and Mechanizing Reason

Humans are tool-makers. We are not unique in this. Rudimentary tool-making and tool-use are known among at least several other species. What has been unique is the extraordinary inventiveness of our tool-making and the range of uses to which we have put our tools. But tools only do so much. They can extend or amplify our own efforts, but they will not do our work for us. Speculatively, it's not much of a stretch to imagine that recognizing the limitations of tools might have spurred the development of draft animal domestication and slavery practices, both of which became common at roughly the same time in the large agricultural societies developing in Mesopotamia some five thousand five hundred years ago.

But while draft animals, slaves, and servants can be made to do tool-using work, they also need to be fed and kept healthy. Resistance and revolt are always possible. So, it's perhaps not surprising that the tool-making imagination would eventually entertain the possibility of creating tool-using beings capable of tirelessly and contently doing one's bidding. The earliest evidence of such an imagined perfection of the tool-making art—the creation of artifacts capable of doing all the work normally undertaken by draft animals and servants—is in the Greek epic, the *Iliad*.[1] There we find brief but tantalizing descriptions of "self-propelled chairs" and "golden attendants" crafted by none other than Hephaistos—the tool-wielding and tool-making god of sculptors and blacksmiths.

Similar stories of artificial servants or companions were fairly common in the works of classical Greek and Roman writers and persisted into early modern times.[2] Often, as in Ovid's *Metamorphoses*, written some two thousand years ago, these artificial beings are humanly crafted statues brought to life by their maker's loving desire and the grace of the gods. Among these stories, *The Fairie Queene*, published by Edmund Spenser in 1590, is unique in featuring an "iron man" that is granted by an immortal to one of the epic poem's protagonists, not to satisfy his personal desires but as a sword-wielding assistant in the noble—if often violent—work of dispensing justice. But in all these premodern tales, even if these artificial beings were crafted by human hands out of clay, stone, or metal, they needed to be animated or "inspired" by the gods.

The first material evidence of imagining that it might be within human reach to build a functioning artificial servant is a set of drawings by Leonard da Vinci. Drafted around 1495, these drawings of a mechanical knight depict inner works comprising an array of pulleys and gears that could be set in motion without divine animation. Interestingly, like the "iron man" who would appear a century later in *The Fairie Queene* and in the dreams of many of those who are funding AI research today, Leonardo's mechanical knight was designed for martial labor.

Human tool-making ingenuity was not up to the task of making anything even remotely like iron men or mechanical knights until well into the last century. Mechanizing mental labor turned out to be much easier than mechanizing physical labor. About the same time Spenser was penning *The Fairie Queene*, a new "curriculum" model of education was being forwarded in which knowledge was a quantifiable good that could be analyzed into component parts for delivery by means of standardized lessons in competitively graded short courses.[3] This new approach to learning was premised on the ideas that reasoning is based on logic and that all forms of knowledge should aspire to the crystalline purity and certainty of mathematical proofs. This association of the commanding heights of human reasoning and intelligence with mathematics—more an exception than a rule across most of human history—proved decisive in setting the course of efforts to build machine intelligence.

By the mid-seventeenth century, machines for performing mathematical operations like addition, multiplication, subtraction, and division were being built in France and Germany, and a "logic demonstrator" was constructed in 1777 by Charles Stanhope (1753–1816) that proved machines could generate logical proofs. Half a century later, after building a "difference engine" that could carry forward the results of a calculation to succeeding operations—an elementary form of machine memory—Charles Babbage (1791–1891) drafted plans for an "analytical engine" that featured a logic unit and an integrated memory, the design of which anticipated the engineering logic and circuitry of the first mainframe computers that were eventually built in the late 1930s.

A major shortcoming of the calculators and logical devices built through the early twentieth century was their reliance on mechanically transmitted energy. Computers made of relatively heavy metal parts require a great deal of energy to set and keep in motion and then suffered from mechanical strain and heat buildup. In effect, the precision limits of machining and assembly effectively set caps on processing speeds and operational complexity. The construction of general purpose programmable computers with substantial working memory became possible only with the inventions of vacuum tubes and solid state electronics that have no moving parts.

Significantly for the course of computing history, it was the paroxysm of the Second World War that put electronic computing on a development fast track. Making advances in weapons design and manufacturing were key military priorities—as were advances in communication and code-breaking—and electronic computers were critical for carrying out the complex calculations involved. Pursuing military/strategic advantage through computational artifice has remained a key driver of basic computing, communications, and AI research ever since.

Modeling Thought: The Research Origins of the Intelligence Revolution

One of the major contributors to Allied efforts to advance computer science was the British mathematician Alan Turing (1912–1954). He was also one of the first scientists to maintain that building artificial general intelligence was within human reach. His core insight was that any act of reasoning that could be converted into a set of algorithms or rule-bound decision-making procedures could be simulated by a sufficiently complex electronic device. While all the operations carried out by such a device could, in principle, be carried out by unassisted human beings at the relatively slow speed of electrochemical exchanges in the brain, the machine's electronic substrate would enable these operations to be carried out at near light speed. Any reasoning that could be formally encoded could also be automated and accelerated.

Over the next decade, remarkably productive mergers were crafted among advances made in what had previously been the largely separate academic research fields of engineering, logic, neurophysiology, evolutionary theory, and cognitive science. By the mid-1950s, the basic principles of cybernetics and the role of feedback mechanisms had been laid out by Norbert Wiener (1894–1964), and growing numbers of mathematicians and computer scientists were beginning to wonder how to best approach building a general purpose AI. A high-level seminar on machine learning was hosted in Los Angeles in 1955, followed by a profoundly influential summer research program on AI at Dartmouth College in 1956, and by a 1958 conference on the mechanization of thought processes hosted at the National Physical Laboratory of the United Kingdom.[4] With these conferences, the Intelligence Revolution can be said to have begun in earnest. AI was no longer seen as the stuff of dreams but as a research agenda worthy of substantial, dedicated investment.

Over the first generation of serious AI research, two major approaches emerged. One approach, building on presumptions about the close relationship among mathematics, logic, and reasoning, was "neat" in the sense that it aimed for precision in programming and in solving well-defined problems. The successes were striking. Digitally computable programs were written that were able to generate proofs for algebra word problems and mathematical theorems, sometimes doing so more elegantly than had previously been done by humans. Systems were built that automatically produced relatively crude but still functional and cost-effective, translations between natural languages like Russian and English. And machine vision and robotics developed to the point that artificial "agents" could carry out simple building procedures and navigate through obstacles in a controlled environment.

But as impressive as the advances made by "neat" research were, with the exception of translation machines and search engines operating on unique databases, the AI that was resulting could easily be dismissed as capable of handling nothing more than "toy" problems. As the philosopher Hubert Dreyfus (1972) pointed out, this was partly because those working on AI drew inspiration from neuroscience and biology regarding how electrochemically stimulated neural networks work in the brain, but then effectively disembodied those networks to create the equivalent of "brains in a vat." Human intelligence, Dreyfus argued, is not primarily a function of symbol processing but rather of embodied interactions in unpredictably changing environments, based on provisional knowledge that is always open to revision.

In keeping with this argument, a "scruffy" research agenda emerged that explored the use of "frames" or "scripts" containing contextual or background knowledge to enable AI to connect with the real world of human activity and not just "toy" problems. This kind of background knowledge is required, for example, to make sense out of the process of ordering from a menu in a restaurant, including how menus are organized, how orders are placed, how much is typically ordered, the meaning of the numbers after each menu entry, and so on. This approach to building machine intelligence subordinated symbolic logic to tacit knowledge or "world models" that an artificial agent could combine with real-world input and feedback to learn how to carry out situation-specific actions.

Throughout this period, from the 1950s to the early 1970s, considerable effort also went into drawing on neuroscientific insights about perception and cognition to build artificial neural networks—interconnected and layered sets of electronic nodes—capable of learning or progressively improving their performance on tasks by extrapolating from a given set of examples, typically without being provided with any rule-based or task-specific programming in advance. By subjecting a large number of photographs labeled as "cat" to multiple layers of progressively refined feature analysis, for example, an artificial neural net will identify consistent feature patterns, strengthening or weighting connections among the "neurons" involved. After this training process, the neural net is then able to examine a random collection of images and identify which ones are images of "cats." Each time the results of such a search and sort operation are evaluated by humans and fed back through the system, the neural network is able to further refine its ability to correctly identify photos of cats.

The neural network or connectionist approach to AI was enormously promising, but in order to work in anything like real-world situations and time frames, artificial neural networks require both large, labeled data sets and considerable amounts of computer memory and computational power. Until those were available, neural networks were destined to remain laboratory curiosities with little real-world application. Neural networks went into what amounted to scientific hibernation.

The AI Investment Winter and Its Aftermath

The continued disparity between projected and actual achievements, across research approaches and agendas, eventually precipitated a sharp rollback of funding for basic and essentially open-ended AI research. In the mid-1970s, the Defense Advanced Research Projects Agency (DARPA) dramatically cut its funding to the major centers for AI in the United States, and what funding remained was directed away from basic research to the development of defense- and security-specific applications like the design of autonomous vehicles, battle management systems, and automation-assisted surveillance. Similar cuts in government funding occurred in the UK in light of criticisms that, after decades of effort, computers played pitifully amateur chess and seemed incapable of even such a simple task as face recognition.

As previously noted, part of the problem at the time was hardware. As Hans Moravec described it, the developmental stage of AI at the time was analogous to having built a functional prototype of an aircraft while lacking an engine of sufficient power to generate the lift required for heavier-than-air flight. But misgivings were also emerging within the AI community about research directions. In a prescient ethical criticism, Joseph Weizenbaum (1976) made the case that too much attention was being given to what *could* be done by AI and too little to questioning what *ought* to be done by it. The profound ethical ramifications of AI were being overlooked in the fervent exploration of what was technically possible. This was a particularly powerful observation given that it was occasioned by the developer of the very first conversational robot or "chatbot" watching people interact favorably with software that provided psychotherapeutic advice with no understanding whatsoever of human experience. At some point, Weizenbaum worried, the behavioral successes of AI would result in potentials for seriously undermining the value of human life and experience.

As it happened, relatively little progress was made in machine conversation over the ensuing quarter century. But, in spite of funding cuts, over the late 1970s and early 1980s, new theoretical work on neural networks dovetailed with accelerating improvements in computing hardware; modest progress was made in building functionally autonomous, computer-guided vehicles; and considerable new work was done proving the usefulness of object-oriented programming in writing software for applications in real-world situations. And when the Japanese Ministry of International Trade and Industry declared in 1981 that it was going to invest an unprecedented $850 million dollars in AI and computing research with the aim of developing a "fifth generation" of computers and programs that would be able to converse in natural language, engage in visual learning, and reason at human levels, the funding winter came to an end. In competitive response, the American and British governments quickly rebooted funding for AI research.

The most notable result of this investment thaw was the emergence of a billion-dollar industry in so-called expert systems. Originally developed in the mid-1960s, expert systems combined a detailed "knowledge base" with an "inference engine" designed on the basis of interviews with human experts in fields relevant to decision-making in the target knowledge domain. Early successes in analyzing chemical compounds and matching disease symptoms with antibiotic prescriptions had proved the validity of the concept. But it was not until the 1980s that it became possible to build general-purpose inference engines that could be "fueled" with domain-specific human expertise and large and fluid data sets. This proved to be a remarkably powerful way of addressing a range of business needs, including monitoring and managing inventory, diagnosing operational bottlenecks, scheduling and guiding equipment maintenance, and evaluating credit applications.

Yet, even well-designed expert systems were susceptible to breaking down when given unusual inputs (the "brittleness" problem), and it was difficult to map out in advance all the preconditions involved in successful action in real-world contexts (the "qualification" problem). And while these systems worked well in decision-making contexts where a few hundred inference rules would suffice, in complex contexts where thousands of rules might be needed and/or where a constantly evolving model of the knowledge domain was required, effective and reliable expert systems were much harder to deliver. Moreover, the growth of personal computing in the early 1990s and the development of more intuitive interface architectures fostered growing decentralization of business computing applications, and by mid-decade the boom in expert systems had largely gone bust.

Although the period from the late 1980s into the mid-1990s is sometimes referred to as a "second AI winter," seen another way it was a period of fruitful convergence and commingling among various streams of AI research. Hearkening back to Hubert Dreyfus's argument that human cognition is fundamentally embodied and environmentally situated, an "actionist" or embodied approach to machine intelligence developed around the idea that intelligence is rooted in sensory-motor coupling with an ever-changing world and in a proprioceptive sense of being present in that world. This yielded significant gains in robotic intelligence. Advances were also made in applying new theoretical work on convolutional and recurrent networks, which greatly improved machine learning performance and proved that there were ways of bypassing extensive supervised training while maintaining full functionality. Other conceptual advances included so-called mixture-of-expert architectures, the application of probability theory and decision theory to AI, the use of "fuzzy logic," and the development of evolutionary algorithms that could rewrite themselves in adaptive response to their informational environments.

In short, at a conceptual level, rather than a second "winter," the period from the late 1980s to the late 1990s was perhaps something more like a protracted spring "cold snap" with lots of new growth going on just out of sight. In retrospect, it is easy to see that what was preventing AI from really coming into its own was not a dearth of innovative science and engineering but a sufficiently rich information environment—an environment with enough data radiance to nurture and sustain the practical embodiment of machine intelligence. That, however, was just around the corner. The information transmission and generation grid known as the internet was scaling rapidly up from being a network used by military and academic elites into a general purpose infrastructure capable of mediating the mutual adaptation of machine and human intelligences in the complex and diverse informational domains of economic, social, political, and cultural conduct.

A New Informational Infrastructure: The Internet, Personal Computer, and Smartphone

The basic design for the transmission grid of this new infrastructure had been commissioned by DARPA and launched in 1969 as the Advanced Research Projects Agency Network or ARPANET. The original motivation for building this "packet switching" system was to have a secure, decentralized, and node-to-node communication network that could withstand nuclear weapons assault—a system capable of sustaining military and governmental communications under worst-case scenarios. It did not take long, however, for the broader potentials of this network architecture to be realized. Connection to ARPANET grew rapidly among American universities and defense agencies and contractors, and international links were established in 1973. A year later, the "internet" was born with the formation of Telnet, the first commercial internet service provider (ISP).

Other ISPs quickly followed. But for the next fifteen years, the internet was still largely used to connect universities, research centers, and governmental agencies. Readily accessible, commercial dialup internet service was launched in 1989, and in the following year, the basic language and text transfer protocols used in developing websites as we now know them were invented (the hypertext markup language or HTML and the hypertext transfer protocol or HTTP). Two years later, the World Wide

Web was inaugurated as a truly public space when the first open-access web servers were turned on.

Web browsers came onto the market over the next few years, along with the first websites for selling goods over the internet, including Amazon and eBay. High speed cable access to the internet became commercially available in 1996. Google was launched in 1998 and its algorithmic search engine quickly became the most widely engaged machine intelligence in the world. To give a sense of the rapidity of the changes taking place at the time, in just the eighteen-month period from December 1998 to August 2000, the number of households in the United States with personal computers rose from 42 percent to 51 percent and the number with internet access nearly doubled from 26 percent to 42 percent. Roughly 80 percent of all children in the United States were suddenly using computers at school, and 20 percent of all Americans were accessing daily news online (US Census Special Report, 2000). While questions continued to be asked about whether the World Wide Web would ever become an environment suitable for profitable commercial activity, at the close of the 1990s, the internet was unquestionably established as a crucial and expanding dimension of the communication infrastructure for the twenty-first century.

The increasing power and decreasing size of microprocessors that were crucial to the personal computer revolution were at the same time enabling both the miniaturization and expanded functionality of mobile communications devices. With the 2002 rollout of the Blackberry, a handheld, internet-linked device with a small but functional keyboard for composing text messages, the era of 24/7 email and internet access was born. The introduction of the first Apple iPhone in 2007—which featured touch screen operation and support for both Web 2.0 (user-generated internet content) and third-party applications—revolutionized mobile communications and triggered dramatic growth in the variety and use of digital social media.

Seen at the level of fiber-optic cables, satellites, and server farms, the phenomenal growth of the digital network infrastructure of the internet was a triumph of physical engineering. But what this physical network made possible was a networking of machine and human intelligences in digital environments that fostered a coevolutionary intelligence explosion. With vast troves of data and aided by spectacular gains in computing speed and memory, artificial neural networks and machine learning algorithms were suddenly poised for unprecedented successes. Over a handful of years, after nearly half a century of concerted effort and slow progress, machine vision and speech recognition suddenly improved to levels that first rivaled and then surpassed human capabilities. From solving "toy" problems and learning to play games like checkers and chess, AI was suddenly able to "graduate," leaving laboratory "schools" to start real on-the-job training. Artificial servants, savants, seers, and soldiers were no longer merely the stuff of dreams.

Artificial Agency and the Goal of Intentional Partnership

Like the internet, the idea of building digitally embodied forms of agency that blend deep machine learning, unlimited information reach, and a natural language interface began as a DARPA brainchild. Although it received almost no media coverage, in 2003, DARPA initiated a project aimed at developing a Cognitive Agent that Learns and Organizes (CALO). It was the largest single AI project that had ever been funded: a $150 million dollar, five-year effort that involved some 350 people at SRI International, a leading technology research and development corporation associated with Stanford University.

The details of the project are instructive. DARPA's mandate was ambitious: develop a personal assistant that could learn onsite and in real time to assist military personnel execute their duties across a range of activity domains from supply management to command post. Such an assistant would incorporate the decision-recommendation capabilities of expert systems, the search and learning capabilities of software agents based on neural networks akin to those in Deep Blue (the computer system that defeated world chess champion Gary Kasparov in 1997), the connectivity needed to carry out commands in a full spectrum of real-world environments, and natural language processing abilities sufficiently advanced to allow completely hands-free partnership. In short, the virtual personal assistant sought by DARPA was one that could be seamlessly integrated into the military workplace and that could not only learn to provide requested information but also to anticipate what information might be relevant and when, offering decision options, managing routine tasks, and carrying out user commands immediately.

According to the SRI website, the aim of the CALO project was "to create cognitive software systems ... that can reason, learn from experience, be told what to do, explain what they are doing, reflect on their experience, and respond robustly to surprise."[5] It was successful enough that SRI exercised its legal right to pursue further research aimed at building a virtual assistant that could be marketed to the public. The commercial potential was obvious. SRI spun off a separate company in 2007 to continue working on a commercially viable personal virtual assistant. Two years later, Siri Incorporated, this spinoff company, premiered a virtual assistant software that could be installed on any smartphone. Within a matter of months, the company was purchased by Apple for an amount estimated to be in the neighborhood of $200 million.

As it had been developed at Siri Incorporated, the virtual personal assistant was envisioned as a comprehensive "do engine" and not just a "search engine." Linked to a suite of more than forty web services, it could suggest alternative travel plans in the case of a cancelled flight, make rental car reservations, book tables at restaurants based on user-stipulated preferences in consultation with a range of restaurant rating sites, and pull together a list of news stories on topics of personal interest in the last ten days. Unlike expert systems that relied on a structured database of knowledge, the virtual assistant could draw on a range of internet-accessible databases and learn how to deploy information from them in completely uncontrolled and unstructured environments.

In 2011, with considerable fanfare, Apple launched Siri—a stripped down version of the virtual assistant with no "do engine" functionality—as a key feature of its new iPhone 4. In the years since, Apple has incrementally added capabilities to and refined Siri, and a host of other virtual personal assistants have come on the market. Most of these have been general purpose "chatbots" that are low on "do engine" or execution capabilities—Amazon's Alexa, Microsoft's Cortana, and Google's Now being among the most well-known. Other virtual assistants have been designed with greater action-capability but for use in specific contexts—for example, travel-related services.

This is changing. Disenchanted with Apple's decision to market a dumbed-down version of the assistant that they had developed, roughly a third of the original Siri team left Apple in 2011 to form Viv Incorporated. Their mission was simple but enterprisingly visionary: make AI a "utility" like water or electricity—a necessity of daily life in the twenty-first century. After five years of development, the company launched its new virtual assistant in spring 2016. According its website, "Viv is an artificial intelligence platform that enables developers to distribute their products through an intelligent, conversational interface. It's the simplest way for the world to interact with devices, services and things everywhere. Viv is taught by the world, knows more than it is taught, and learns every day."[6]

Unlike expert systems, Viv is not programmed to perform specific tasks or supplied with a fixed knowledge base. Armed with state-of-the-art natural language processing and constructed around the deep learning architecture behind the successes of Deep Blue and Alpha Star (a computer system that learned on its own how to play the multiplayer strategic game StarCraft II at grandmaster level), Viv is able to interpret a user's intention and to write a program for executing that intention by assembling all the required resources from as many different digital and real environments as necessary. And it is able to do this in a matter of milliseconds. One might inform Viv, for example, that "It's my one-year anniversary with my girlfriend. Have a dozen red roses delivered to her apartment this morning with the message 'It's been the best year of my life; dress nice; dinner at 6.' Have a car sent to pick her up so she's sure to arrive at Mario's on time. If it's still raining at the end of the day, get a car for me, too. Let the manager know that we'll be having the stuffed whole snapper and to pick a nice white wine." Viv will carry out your instructions and be ready to make adjustments on the fly, getting the car to arrive earlier, for example, if traffic turns out to be particularly bad.

The working premise of Viv is that the only limit to its capabilities will be the number of service, product, or knowledge providers who are willing to link into its network. Unlike Apple's Siri, Viv is designed to be available through any device incorporating a microphone component, from smartphones to cars to tablet computers or home entertainment systems and refrigerators. The company's hope is that tens of thousands of service providers will link to the network, turning it into a "global brain" dedicated to making commerce conversational.

The advent of artificial systems of agency like Viv is almost certain to be epoch-making. What we are seeing is the transition from digital machines that execute commands stated in rigorous digital code to machines that are capable of independently enacting human intentions expressed in natural language. This "intentional partnership" makes it possible for our personal agency to become distributed—enabling us, in effect, to be actively present in many places at once. Virtual personal assistants can set events in motion anywhere that is internet-connected, moving real goods and establishing service relationships at near light speed, and, as they do so, they will be training themselves to think strategically on our behalf, recommending some courses of action and questioning the wisdom of others. With almost no media fanfare, we are entering an era of what amounts to human-machine symbiosis and karmic partnership.

The Role of Big Data

In theory, there is no limit to what algorithmic intelligence can learn to do.[7] In practice, however, algorithm-based machine learning is a very data-hungry process. The precise relationship among data, information, and knowledge is open to debate, but the general consensus is that data is foundational. Data is what is "given" or what is often thought of as the raw facts. But to fully appreciate how dramatically the human-technology-world relationship is being transformed by intelligent technology, it is important to be clear that data is not value neutral. Every piece of data (that is, every datum) is a sensory/ observational atom that consists in a measured value for some specified variable. For example, the level of mercury in a thermometer provides a measured value for the variable of temperature. If the thermometer was just removed from my mouth and the mercury stands at 102, this observed data can be interpreted as information that I am running a fever.

As this example illustrates, data are not so much raw facts *about* the world as they are targeted records of occurrences *in* the world. Thus, when we talk about algorithms combing through data for patterns, they are combing through the results of recordings of events that in some way have mattered to humans. The datasphere is a space of all the records of events that we humans have for some reason deemed worthy of attention and recollection. Data are memory traces of human intelligence in action. Thus, although we use the term "artificial intelligence" to refer to what is being demonstrated by virtual personal assistants, by game-playing software systems, and by online search and recommendation services, what is actually being demonstrated is a form of synthetic intelligence—a blending of artificial and human systems for doing things like recognizing patterns and making judgments.

Prior to the late 1990s, while the internet and the World Wide Web were rapidly expanding in size, they were not actually generating huge amounts of new data. They were simply expanding the number of connection points from which existing bodies of data could be accessed. This changed with the advent of 24/7 connectivity, social media, e-commerce, and the efforts of surveilling governments and corporations to make data-generation and data-sharing as natural and desirable as drinking water and breathing air. In the space of just a few years, and with open-ended evolutionary potential, data-hungry synthetic intelligences gained access to virtually limitless nutrition.

To give a sense of the scale shift in data production, while 100 gigabytes of data were produced globally per *hour* in 1997, five years later that same 100 gigabytes of new data were being produced every *second*. By the end of 2017, roughly 50,000 gigabytes of new data were being generated every second, including roughly 204 million emails, 216,000 Instagram posts, 277,000 Twitter posts, and 72 hours of YouTube videos.[8] In the first quarter of 2019, internet users were generating over 2,500, 000,000,000,000,000 bytes of new data every 24 hours—the equivalent of 10 million high definition video discs. If that is not astounding enough, consider that the world's 4.54 billion internet users and 3.8 billion social media users spend an average of 6 hours and 42 minutes online each day. In other words, over the course of this year, humanity will spend a cumulative *1.25 billion years online.*[9] Data production and sharing have not only been "democratized," they have become benchmarks of digital social normalcy.

Yet, these are only the most visible aspects of big data. The incorporation of miniature, networked sensors into everyday objects is infusing data production capabilities into the things we interact with daily, ranging from our pens, running shoes, and medicine bottles to our cars and our refrigerators. This infusion of connectivity into everyday objects—the creation of the so-called internet of things—is producing a world of ever more enticingly "enchanted objects" (Rose 2015). Running shoes embedded with internet-connected sensors keep track of your pace and the routes you run and calculate how many calories you burn. The caps of your elderly parents' "enchanted" medicine bottles change color to remind them when they need to take their medications, while at the same time sending records of whether they are doing so to their health care providers. It's estimated that by 2025, the average, connected person will interact with some 4,800 such devices per day and that this will result in the generation of 163 zettabytes of data globally per year (Reinsel, et.al. 2017). To put this in perspective, in a single year, humanity will produce enough data to make a high definition video lasting longer than the five-billion-year history of the Earth.

Drawing on data gathered from credit/debit card purchases, web searches, and text, image, and video postings to social media, algorithmic intelligences have become remarkably effective at personalizing product and service advertising and pricing. As might be imagined, this is a very valuable skill.[10] But, in addition to running recommendation engines, algorithms nourished by big data are also skilled at producing consumer credit ratings; organizing airline flight schedules; making "risk assessments" and "evidenced-based" recommendations regarding bail, sentencing, and parole; and finding patterns of disease treatment effectiveness that have until now eluded human recognition.

All of the data that is now being uploaded into the "cloud" by our use of internet service providers, social media platforms, online retailing, credit/debit cards, smartphone payment apps, navigation devices, and the internet of things also falls as data "rain" that can be channeled back into AI development and "deep learning," further energizing and extending the reach and effectiveness of algorithmic agency. In short, the more we make use of virtual assistants and deploy algorithmic agencies, the more transparently and powerfully they will be able to respond to our expressed needs and desires, but also the more precisely they will be able to interpret and anticipate our actions and intentions.

The Fourth Industrial Revolution: A Revolution in the Cloud

The confluence of evolutionary machine learning algorithms and big data is changing both the pace and character of the Intelligence Revolution.

Although it is still apt to regard it as an industrial revolution, it is industrial in a new way. Asked to think about an industrial revolution, although we know that the factories of today are not like their nineteenth and twentieth century forebears, most of us will still envision blunt-faced, utilitarian structures built of brick or concrete, bustling with activity and exhaling plumes of fossil fuel smoke. Inside them, we would expect to find skeletal constructs of steel and brass set into cacophonous motion by steam engines or electrical generators, devouring raw materials and step-by-step transforming them into conveyor belt-conducted parades of identical finished products. These are valid imaginations. Until quite recently, most industrial production was carried out according to a marvelously visible logic of moving mechanical parts. You could see it taking place.

Asked to reflect on the presence of AI, machine learning, or big data in our lives, most of us will think first about computers and smartphones and the recommendation "engines" and navigation services we access through them. Or, we might think about self-driving cars or robotic surgeons. In short, we are inclined to think in modern industrial terms about material objects and processes: physical machines that somehow manage to behave intelligently. The "factories," "machines," and "products" specific to the Intelligence Revolution, however, are neither strictly located nor directly visible. It is true that there are data server farms and cloud computing campuses housed in structures that can be as large as six million square feet in floor area and that individually can require millions of gallons of water daily for cooling purposes. But these buildings and the equipment in them are not the actual factories of the emerging AI industries. They are analogous at best to the brick and mortar shells in which factory equipment was operated in the heyday of the Machine Age, many of which have now been gainfully repurposed as innovation centers or commercial complexes.

Walking along a server farm's seemingly interminable, identical aisles of head-high racks of lightly glowing equipment faces, you will not *see* any of the work being done. The Intelligence Revolution is industrial, but its factories and machines are computational. They are not built out of concrete or metal but out of mathematical and logical codes. It is these codes that "magically" instruct the movement of electrical energy through circuits so finely etched into silicon substrates that more than twenty-five million transistors can be fitted into a single square millimeter. There are causal processes at work, but we are not in a position to physically witness them. Open up a smartphone or a tablet computer and you will not see machine intelligence at work. You will not witness the transformation of any raw materials into finished products. The industrial factories and machines proper to the Intelligence Revolution exist, but they do so in the way that ideas exist—without a specifiable spatial location.

This fact is obscured by our fascination with things. Although we now refer to internet-enabled things as "smart," they in fact only transmit intelligence. Disconnected from the internet, smartphones and smartspeakers are no more intelligent than old rotary dial phones and transistor radios. They are tools that, on their own, are incapable of adaptive conduct. Once connected, however, smart devices begin functioning as *iconic* artifacts, mediating our attentive immersion in information and communication technology, transmitting our wishes, our likes and dislikes, our judgments, and our dreams into algorithmic factories where they are analyzed and fashioned by the computational analogues of heavy equipment. Like the bulldozers and cranes of the modern era, this computational heavy equipment is being used to refashion the human environment, building new informational infrastructures, constructing virtual megacities out of nothing more material than human desires, purposes, and meanings.

The factories of the industrial revolution in the nineteenth and twentieth centuries facilitated a massive concentration of human populations and energies in sky-scraping urban environments. The results of that technological experiment have been mixed: a volatile stew of previously unimaginable human creativity and equally unimaginable waste and squalor; a world of climate disruption and environmentally forced mass migrations; a world in which the human and the natural are no longer coextensive; a world that can be described either as one in which nature has been dehumanized or as one in which the human has been denaturalized.

The factories of the Intelligence Revolution are facilitating a similar concentration of populations and energies but in cloud-computed virtual metropolises answering human needs and desires without ever once involving direct human touch. The results thus far seem likewise to be mixed. With the advent of cloud-based virtual assistance and the internet of things, the presence of AI can no longer be thought of as entity-like. This is not an intelligence that is constrained by anything like an individuated (computational or robotic) existence. Rather, it is an intelligence that is being infused into the human experience as an *ambient presence* in our lives—an environment of mutual adaptation and coevolution in which the "climate" is a function of changing expressions of our own likes, dislikes, values, and desires. It is an intelligence that is becoming dynamically *coincident* with our own.

Given humanity's failure thus far to effectively resolve global predicaments like climate change and the persistence of hunger in a world of food excess, this dynamic coincidence of human intelligence and AI should concern us.

Entering into symbiotic relationships with AIs dedicated to anticipating and gratifying our individual desires will not only result in us getting what we want more swiftly and surely, it will also reinforce our wants and desires and intensify any conflicts among them. The Intelligence Revolution has brought us to the verge of a historical and karmic turning point.

3

Intelligent Technology: A Revolution in the Making

Concerted efforts to build intelligent machines have been ongoing for nearly seventy years. Informed by a wide range of scientific and engineering strategies, these efforts are now succeeding in bearing significant and remarkably varied practical fruit. The aim of this chapter is to situate these successes in their broader social, economic, and political contexts and to examine how intelligent technology is currently reshaping the human experience.

Before doing so, however, it is useful to clarify the meanings of both intelligence and technology. One of the implications of the Buddhist teachings of interdependence and emptiness is that we should be wary of assuming that we have common conceptions of even such apparently straightforward elements of daily life as eating utensils, much less such complex phenomena as intelligence and technology. While "eating utensils" can readily be translated into Japanese, the actionable possibilities implied are very different. The most common eating utensils in Japan are not forks and spoons but foot-long, tapered sticks of lacquered wood called *hashi*, and eating in Japan includes the possibility of consuming (and relishing) *natto*, a slimy, fermented soy bean product that most Americans would refuse to consider "edible." The relational scopes of "eating" and "utensils" are not common across cultures. Likewise, there are no universal sets of actionable relational possibilities associated with intelligence and technology.

Intelligence: A Working Definition

A fairly standard ensemble of traits associated with intelligence are the abilities to learn, to make reason-based judgments, and to identify goals and act as needed to achieve them. This ensemble of traits would at present preclude speaking about artificial or machine intelligence in anything but a speculative manner. Machines do not yet set their own goals or formulate intentions. A definition friendlier to those wishing to entertain the possibility of intelligent machines is "an ability to accomplish complex goals" (Tegmark 2017: 50). According to this minimalist definition of intelligence, so-called smart systems like those used in online recommendation engines are behaving intelligently; they are accomplishing the complex goal of inducing consumers to purchase specific goods or services. This definition seems overly broad, however, since its agnosticism about the origin of goals suggests that unlearned and exogenously determined behaviors, like those of an air-conditioner thermostat, can count as intelligent.

For our purposes, I would like to identify intelligence with adaptive conduct. This links intelligence to both learning and goal orientation but not necessarily to intentionality or subjective experience. "Adaptation" implies both permutation and persistence in a changing context, while "conduct"—which can be traced etymologically to the Latin *conducere* or to "bring together"—stresses relationality or mutually engaged action. Adaptive conduct thus consists in recursively causal *interactions* of actors and things acted upon. Put somewhat differently, intelligence emerges and is both embedded and shaped in relation to a dynamically responsive environment of actionable possibilities. Intelligence implies active coordination.

This way of conceiving intelligence allows us to regard all living things as intelligent. Microorganisms, some so small that 150,000 can fit atop a single human hair, have minimal capacities for sensing and responding to changes in their environments. Yet, they have proven adaptive enough to be able to survive in environments as different as alpine glaciers and undersea volcanic vents. The same is true of the largest living things, including giant sequoias and single-entity fungal colonies that can cover several square miles. Again, although lacking brains and presumably any subjective experience, these species have proven their adaptive capabilities at evolutionary time scales. They exhibit what we might call "slow" intelligence.

Animals display similar "slow" or evolutionary intelligence. But animals generally are also able to respond as individuals to changes in their immediate circumstances and to learn from doing so.[1] Unlike microorganisms and plants that have very limited environmental horizons, the capacities for sensory coupling enjoyed by animals are much more varied. As a result, their lived environments and potentials for adaptive conduct are much less constrained. Some domesticated animals, especially pets, engage in inference-embodying conduct aimed at influencing—and not merely reacting to—their environments. A dog, for example, might learn to run to the door and bark in order to get its owner to vacate the chair it wants to occupy, eliciting from the owner a laughter-laced scolding and a snuggle on his or her return. In this case, the "fast" intelligences of both the pet and the owner are linked in ways that go beyond providing evidence of their independent goal orientations and offer, as well, evidence of goal coordination and potentials for the emergence of truly shared goals.[2]

In addition to allowing intelligence to be seen at work in all living organisms, this definition also allows us to entertain the intelligence of nonbiological systems that respond adaptively to changes in the environments relevant to their own persistence. The intelligence evidenced by artificial systems, as with organic systems, will be relative to their capacities for sensory coupling and for affecting their environments. A "smart thermostat" like Google's Nest learns what temperatures you like at various times of day rather than initiating heating or cooling when preprogrammed temperature thresholds are reached. Possibilities thus exist for the emergence of many different "species" of artificial and synthetic intelligence. And with sufficiently complex data input and processing, there is no reason why informational machines cannot display both quite broad and flexible intelligence. The simple fact that the material substrate of these systems is silicon-based circuitry rather than carbon-based biology does not preclude them in principle from engaging in highly complex adaptive conduct.[3] Indeed, since artificial intelligence (AI) could enjoy the full range of sensory coupling and response capabilities as organic systems, supplemented with capabilities for coupling, for example, to statistical or digital network environments, the actionable potentials of AI could in theory exceed those of any organic system. Machines, like humans and other animals, are capable of exhibiting both slow and fast intelligence.

If there is a plausible line of demarcation between the adaptive conduct of biological and artificial systems, it is with respect to their goals. Metaphysical questions can, of course, be raised about whether biological systems, including humans, have any truly endogenous or internally generated goals. Serious arguments have been made by materialists that we believe we possess a high degree of freedom in our goal orientations only because we lack sufficient understanding of the deterministic laws of nature. No less serious and opposing arguments have been made for seeing both life and the possession of free will as gifts of a transcendent creator, without which the universe would be a pointless swirl of dead matter. Happily, we do not need to settle or take sides on these long-standing metaphysical debates. The goals of machine intelligences, for now and into the near-term future, must be given by humans.

It may one day be possible for artificial systems to generate their own goals, enjoying the same (real or illusory) freedoms of choice and orientation as we humans do. But until then, all machine goals will originate exogenously with human designers and programmers. Concerns about whether the conduct of artificial agents might become misaligned with human values are at best premature. For the foreseeable future, the goals and values of AI(s) will be in precise and unfaltering alignment with human interests and values. And *that* should concern us. The benefits of intelligence depend on the purposes to which it is devoted.

The Nonidentity of Tools and Technologies: A Critical Wedge

Technology is often presumptively claimed to be value neutral. The corollary of this is that technologies have moral valence only in terms of how they are used by humans. Hence, the proclamation of gun rights advocates that "guns don't kill, people do." Since machine intelligences are not generating their own values and intentions, the moral valence of intelligent technology is limited to its uses and misuses by humans. These claims seem quite plausible. But their plausibility depends on a kind of philosophical sleight-of-hand—a misdirection of critical attention from technologies to tools that renders invisible the distinctive risks and ethical challenges posed by intelligent technology.

Tools are crafted to amplify, extend, augment or otherwise alter our capacities for carrying out particular kinds of work. Tools can be refined or refurbished to enhance their task-specific utility, and they are aptly evaluated in terms of how adequately and accurately they enable us to accomplish what we want. That is, tools exist solely as means and do not specify the ends to which they are employed. In addition, no matter how widespread the use of a particular tool might be, whether we make use of it personally is for each of us to determine. With tools, we can always exercise "exit rights." We can opt out of taking advantage of the ways they would extend our capabilities for acting in/on the world and with/on one another.

The same is not true of technologies. Technologies are emergent systems of material and conceptual practices that embody and deploy both strategic and normative values. Technologies qualitatively transform the ways we relate to the world around us and with one another, shaping not only *how* we do things like seeking entertainment and communicating, but *why*. They are systems for scaling up and structuring human intentions in ways that, over time, also change the meaning and nature of both human intentions and intentionality. Technologies cannot be value neutral.[4]

Whereas tools are built and localizable, even tools as extensive as nationwide electrical grids, this is not true of technologies. Technologies emerge from and subsequently inform and structure our conduct in the same way that natural ecosystems emerge from and then dynamically inform and structure species relationships. We do not *build* or *use* technologies, we *participate* in them. Technologies are relational media within and through which we exercise and extend the reach of our intelligence in pursuing what we value. They are dynamic *marriages of structure and agency* and as such cannot be evaluated in terms of task-specific utilities. Technologies can only be evaluated *ethically* in terms of how they qualitatively affect human–human and human–world relational dynamics.

If we return to the "guns don't kill" argument, while it is true that, like other tools, guns are incapable of making murderous mayhem unless turned to that task by human hands, it is no less true that weapons technology scales and structures human intentions to inflict harm while making apparent the value of doing so from as great a distance as possible. Developing and deploying weapons technology thus comes with structural risks associated with transforming the meanings and stakes of both attack and defense.

As long as attention is directed away from weapons technology to guns, it seems perfectly reasonable to address the risk of children accidentally harming themselves while playing with loaded guns or the risk of violent crimes being committed with illegally possessed weapons by ensuring that all guns are built with biometric identification systems that keep them from being used by anyone other than their registered owners and those to whom use rights have been formally extended. Similarly, the risk that social media platforms might be used to foment racist sentiments or that sentencing and parole algorithms might accidentally reinforce patterns of social injustice would seem to be adequately addressed by means of one or another technical fix—rigorously designed systems of human or algorithmic gatekeepers and data set auditors.

But tool fixes, no matter how extensive or ingenious, are impotent with respect to the risks involved in reconfiguring human–human and human–world interdependencies. Technical solutions to "problems" like gun violence involve building better and safer tools and thus placing constraints on acceptable utility. They are responses to unwelcome events that address the final causal phase when an agent and a tool-enabled action inflict harm. If technologies are values-infused decision-making environments that are structured by and in turn restructure human agency and values, and if the human-technology-world relationship is one of reflexively complex network causalities, then technological harms are not *preidentifiable events*; they are predicament-laden *emergent phenomena* that grow recursively over time, becoming part of the relational fabric of practical decision-making.

Owning a gun is one response to the risk of unwanted intrusions and threats to one's property or one's person. As a personal security tool, kept safely in one's home, one's gun clearly has nothing to do with school shootings or armed robberies. But mass shootings and armed robberies are not intrinsically aberrant uses of guns, and gun violence does not occur in psychic, social, economic, or political vacuums. Increased *readiness* to inflict harm at a distance is a risk of weapons technology. The willingness to trade data privacy for connective freedoms of choice is, we will see, a predicament-expressing structural risk of intelligent technology and the attention economy in which is it implicated. Understanding and addressing technological risk is not a process of anticipating and guarding against future harmful events; it is a process of *anticipating dynamic relational patterns* that become harmful as they and their informing values are scaled up to become part of the environments in and through which we express what it means to be human.

The Practical Absence of Exit Rights

An important benefit of drawing a clear distinction between tools and technologies is that it enables seeing how we can be affected by technologies even when we do not directly make use of any of the tools associated with them. Once a technology is deployed with sufficient intensity and at sufficient scale, we cease to have substantive exit rights with respect to it.

To take a simple example, I may decide not to use the texting capabilities of my smartphone. But texting affects peoples' planning behaviors, informing their readiness (or reluctance) to make firm commitments. Texting encourages communicative brevity and forfeitures of reflection and has also come to serve as a kind of social sonar through which people constantly establish their presence and assess their personal status. Average Americans now check their phones several hundred times a day, at once locating and distributing themselves in a social space that is "public," but only for those able to access and peer through its digital infrastructure. These changes affect me. In the social world emerging with new communication technologies, not only are my refusals to take advantage of texting and social media networks seen by many people as acts of willful disassociation, these refusals do nothing to halt the technological saturation and transformation of my own, nondigital communicative environment. Whether or not I opt out of using these web-based tools, I cannot avoid being impacted by how their nearly ubiquitous use is transforming, for example, what it means to be a good friend.

To put this somewhat differently, tools and technologies exist at different ontological levels or orders of reality. The information philosopher Luciano Floridi helpfully speaks about this as existing at different "levels of abstraction" (Floridi 2013). Consider, as an analogy, the relationship between chemicals and living organisms. Contemporary biology tells us that life originates in complex chemical interactions and that living organisms can be seen as nothing more than self-replicating chemical machines. But living organisms nevertheless behave in ways that cannot be reduced to or described in purely chemical terms. No matter how detailed an understanding we may have of the chemical bonds incorporated in a tiger, for example, this tells us nothing about the tiger's predatory habits or mating behavior. Tiger behavior is explainable only at a level of abstraction much higher than that of chemical bonds. Technologies are emergent relational systems that can no more be understood or evaluated at the level of abstraction at which individual tool use occurs than the behavior of living beings can be understood in terms of chemical interactions.[5]

Some philosophers of technology have tried to signal this ontological difference by claiming that technology is a complex and ever-dynamic "milieu" in and through which we define ourselves as human (Ellul 1964) or by claiming that "technology is essentially mediation" (Puech 2016: 83). But I think this tacitly attributes to technology an inappropriate passivity. The conscious exertions of a tiger intent on chasing down prey affect chemical processes throughout its sensory-motor system. In Floridi's terms, events at a higher level of abstraction can affect events at a lower level of abstraction. Or more generally stated: the behavior of higher order emergent systems can causally affect behaviors at the level of their constitutive sub-systems. This is what is referred to as "downward causation" (Andersen, et al. 2001). To see technology as a milieu or structure of mediation is to overlook its recursive causal effects on the human practices out of which it has emerged and in interdependence with which it persists.

This recursive causality is what warrants the metaphor of seeing technology as existing environmentally or ecologically. But technologies are not purely or even largely natural or material. What sets technologies apart from natural ecosystems is that they are ecologies of consistently enacted values and intentions—karmic environments that have emerged out of and recursively habituate us to certain patterns of purposeful human activity, stabilizing and amplifying preferred patterns of outcome and opportunity. To identify technologies with the iconic artifacts associated with them is analogous to mistaking *instruments* like a judge's gavel and sounding block and their use to modify courtroom behaviors with the workings of the justice system as a social *institution*.

Intelligence Industries and the New Attention Economy

As an environment of conduct-structuring relational innovation and novel forms of adaptive agency, intelligent technology and its societal ramifications cannot be understood or evaluated solely in terms of our individual uses of smart devices or internet-accessed smart services. This is in part because the impacts of using them go far beyond the horizons of their task-specific utilities. When devices like computers, smartphones, and smartspeakers are connected to the internet—and through it to the kinds of agency manifested by machine intelligences—they cease to operate merely as ready-to-hand tools and take on *iconic* functions.

The purpose of icons is to attract and transmit but not hold attention. Religious icons, for example, are ritual instruments that have sacred value to the extent that they attract and direct our attention from the mundane world to a specific religious beyond. Computers, smartphones, and smartspeakers are instruments that enable us to carry out mundane tasks like contacting friends, purchasing, storing and sharing music and image files, and ordering restaurant food delivery. But, in addition, they serve the iconic function of directing our attention and the data carried along with it into the material infrastructure of digital connectivity and from there into an immaterial realm that is otherwise inaccessible.

Like the "beyond" into which religious icons provide access, the technological realm into which our attention is conducted by smart devices is essentially mysterious—a computational environment configured by invisible and (from most of our perspectives) "supernatural" forces and agencies. In much the same way that religious practitioners direct their appeals for support and guidance to their chosen icons, we faithfully submit our appeals for assistance and guidance to our web browsers and such favored digital "deities" as Google, Amazon, and Alibaba. Then, by means that we neither witness nor comprehend, we are rewarded with helpful connections to information, goods, and services.

The efficacy of religious icons is often said to be proportional to the depth of religious practitioners' faith. It is the fervor of a practitioner's belief that empowers the icon as divine intermediary. There is a similar relationship between us and the icons of the Intelligence Revolution, only the efficacy of these icons is not an obscure function of religious fervor. It is transparently and directly proportional to the depth and duration of the attention we lavish on these icons. Through their virtually seamless merger of ordinary and iconic functions, our instruments of digital connectivity facilitate the computational conversion of attention-transmitted data streams into both revenue and the power to progressively fashion human behavior.

Attention generates and transmits data. Even just momentarily flicking our eyes to an ad while using a camera-equipped, internet-connected device generates data about what we find attractive and for how long. The fluctuating rates at which we scroll through a menu of YouTube or Netflix offerings produce data about subliminal considerations that may contrast sharply with what we consciously choose to watch. Like radio waves, attention is a carrier of meaning-laden content—in this case, content regarding what is being deemed significant, how it is being responded to, and with what changes over time. In a very real sense, capturing and holding attention is deeply and intimately revealing *intelligence gathering*.

The global infrastructure of digital connectivity makes possible intelligence gathering at industrial scales—a real-time mapping of mercurial shifts in the textures of our personal needs and desires, as well as the more stable complexion of our values and intentions. As we will soon be seeing, this is the deep story behind the explosive growth that has been enjoyed over the last decade and a half by internet access and service providers, social media platforms, and online shopping sites—a story of legal but ethically troubling commercial espionage for the purpose of behavioral prediction and control. But it is first important to be as clear as possible about how attention factors dynamically into the karmic environment emerging with intelligent technology.

We can begin with an analogy. The Earth's ecological dynamics are a complex function of how solar energy is differentially absorbed and transmitted by the atmosphere, oceans, and land masses and how resulting climate phenomena shape and are in turn shaped by biotic agents making intelligent use of this energy. Similarly, the environmental dynamics of intelligent technology are a complex function of how attention energy is absorbed and transmitted through the global technical/material infrastructure of fiber- and satellite-connected computers, smartphones, routers, servers, and data centers and then utilized by intelligent and variously motivated societal agents, thus shaping the human-technology-world relationship.

One dimension of the reshaping of this relationship with the emergence of intelligent technology is the sharp acceleration of ongoing shifts in the logical structure of economic growth and from the relative primacy of material industries to mental industries. Material industries are predicated on real property rights and operating "brick and mortar" factories. Their revenue growth is entrained, first, with increasing volume through accelerating resource acquisition, production, and delivery and, secondly, with the use of their resulting fiscal power to acquire more fixed assets. In mental industries, primacy is accorded to intellectual property rights, and what most powerfully affects revenue growth is success in research labs and design studios. Market value and profitability are not a function of quantities of products manufactured but of qualities of prototypes designed and their effects on patterns of resource acquisition and circulation (Lash 2002).

With the advent of intelligent technology, the most important of these resources has become attention. Although it is still common to refer to the global economy as an information economy, this is no longer particularly apt. If economics is broadly speaking the process by means of which resources are allocated in society, especially scarce resources, then what has been emerging over the last twenty years is not primarily an information economy. Big data has made information abundant and cheap. What is scarce, and hence most valuable, is attention. The logic of an economy of attention was straightforwardly (and presciently) stated by Herbert Simon (1971), who insightfully observed that an abundance of information leads to a scarcity of attention. Central to the character of contemporary global network economics is the intensifying competition among information sources for attention share in the context of a shift from an economic logic of accumulation toward a logic of flows in which *value* is essentially and literally a function of *currency*.

The term "attention economy" has now entered the vocabulary of mainstream business and economics, in the contexts of which it generally refers to the pivotal role of attention capture and retention in generating consumer demand and desire (see, e.g., Davenport and Beck 2002; Brynjolfsson and Oh 2012). Tim Wu's books, *The Master Switch* (2010) and *The Attention Merchants* (2016), present incisive looks at the historical emergence of this aspect of the attention economy.[6] As he very clearly documents, while it is arguable that something like an attention economy has operated in all human societies—quite explicitly in art and antiquarian collecting, and more broadly in uses of political regalia, lavish religious rituals, and even the uses of jewelry and makeup—it was only with the print and broadcast media of the nineteenth and twentieth centuries that it became possible to convert attention into revenue at mass scale. This can be called the Attention Economy 1.0.

The intelligence-driven attention economy of today is something significantly different. It is an economy fueled by industrial extractions and circulations of attention and the data traces of human intelligence carried with it as resource commodities that can be traded freely and individually for ever more briefly satisfying sensory, emotional, and epistemic goods and services. In the Attention Economy 1.0, large population market studies, crudely aggregated price signals, and mass advertising were used to capture attention and stimulate demand for mass-produced goods and services. In the Attention Economy 2.0, the internet functions as a combined communication and production infrastructure through which users of digital search, social media, and e-commerce are drafted into double duty: as *consumers* of individually targeted material and informational goods and services and as *producers* of highly granular training data for "smart" systems laboring tirelessly and creatively to accelerate and expand revenue-generating processes of attention capture and exploitation.

This new attention economy is without doubt a complex and logically beautiful system. In a material economy, accelerating consumption accelerates resource use and depletion. Even if the resources used in producing and delivering material goods are relatively cheap—for example, lumber and labor in the manufacture of furniture—they come at a cost. In the emerging immaterial economy of attention, every act of consumption is simultaneously an act of cost-free resource provision and procurement. In addition to any monetary returns received by those involved in digital advertising and in producing and delivering internet-ordered goods and services, they are also receiving data returns that can be sold or traded in addition to being used recursively to adaptively revise further product design and marketing. In short, the infrastructure that enables digital commercial actors to respond to consumer desires serves simultaneously to gather intelligence that enables these actors to produce with ever greater precision the kinds of consumers and desires that will be most profitable for them. The Attention Economy 2.0 is a complex computational system through which the apparently distinct processes of articulating, granting, and generating further wishes are folded into the dynamic equivalent of a Klein bottle—a three-dimensional construct that has neither inside nor outside.[7]

It is crucial to appreciate the fact that while the Attention Economy 2.0 functions as a perpetual desire machine, it also accelerates the consumption of both material and immaterial goods and services. Attention captured is time consumed. Time consumed is not time invested. Structurally, the intelligent attention economy is a system for transmuting freely given time into needs for both material and immaterial goods and services that could otherwise have been met directly by those whose time has been consumed in exploring digital and analogue environments which have been constructed for that very purpose. The economically critical factor here is not the content of the media consumed or the informational searches and product purchases made online; it is the value of time/attention that could have been otherwise spent or invested.

A middling global number for per capita mass media consumption and internet use is just over six hours per day. Multiply this by a world population of (say, five billion) active learners and laborers and this works out to thirty billion hours of daily "trade" in attention. Multiply again by the investment value of this time—if used to engage, for instance, in productive activity like growing or preparing food, maintaining homes or vehicles, caring for family members or neighbors, learning new practical or aesthetic skills, or pursuing formal education—calculated according to a very modest purchasing power parity estimate of five dollars for an average hour of attention. This works out to a global daily transfer of $150 billion dollars of time and attention investment capital and an annual total of roughly $54 trillion. The world economy, as calculated by the World Bank for 2019, was just under $90 trillion.

The value of the intelligence gathered by network sirens and solicitors is well reflected in the global ranking of leaders in the intelligence industries. What is not well reflected is the commercial value of the lost *investment potential of time and attention capital* that is being freely transferred from the world's people to these industries. What might we accomplish if we collectively invested an additional thirty billion hours of time and attention daily in actively and directly contributing to the lives of those around us? How much more valuably would we find ourselves situated if our daily global total of $150 billion of time and attention capital was invested in enhancing our domestic environments and our neighborhoods and their connections to the natural ecologies that support them? Families and neighborhoods are like plants. If they are nourished with time and attention, they thrive. If not, they fail to grow, atrophy, and perish. What changing patterns of outcome and opportunity are resulting from this mass transfer of attention capital, and are there any grounds for seeing them, not merely as good for some but as superlative for most?

At present, the patterns of outcomes and opportunities that are emerging and being consolidated by intelligent technology are consistent with it functioning as an environment conducive to scaling up and marrying three sets of intelligence interests—commercial, political, and military. Like our own marital institutions that vary culturally and over time, these marriages of intelligence interests are neither identical nor static. Neither are they necessarily or even ideally marriages among equal partners. What they share, however, are the karmically charged values of predictive precision and control.

When Charles Babbage built his "difference engine" almost two hundred years ago, anticipating many of the features of today's digital computers, his aim was to produce flawless logarithmic tables for use by engineers and navigators. These tables previously had been computed by hand, and because the output of calculations performed at one level of such tables feed into the calculations performed at subsequent levels of the table, mistakes were self-amplifying. Using an error-ridden table could have disastrous, real-world consequences, and a significant part of Babbage's motivation for building the difference engine and designing his analytical engine was to bring about a safer and more certain world.

The informational engines of the Intelligence Revolution are similarly being designed to reduce error and refine or expand capacities for real-world control. The development of autonomous vehicles, for example, has the potential to reduce accidents and increase safety. The use of algorithmic engines to produce data-driven recommendations on bail, sentencing, and probation can help eliminate human bias and the effects of judicial staff fatigue, while at the same time improving judicial efficiency. But unlike Babbage's difference engine, the algorithmic systems embedded in autonomous vehicles and judicial support systems are not just carrying out mathematical calculations; they are assuming many of the roles performed only by human navigators and engineers in Babbage's day, processing real-world data to make decisions and to recommend courses of action. Machine intelligences today are not merely interpolating mathematical values, they are interpolating and interpreting human values, generating ever more minutely accurate tabulations of human propensities and desires, often with competing purposes for modeling, predicting, and shaping human decision-making and relational dynamics. This raises two important questions. To whose benefit is this being done? And to what ends?

Commercial Intelligence Interests

Tim Cook, writing as the CEO of Apple in 2016, made the very revealing statement that "when an online service is free, you're not the customer. You're the product" (Cook 2016). This echoes an almost identical and more specific statement that was made by the cultural historian and media scholar Siva Vaidhyanathan: "We are not Google's customers; we are its product" (Vaidhyanathan 2011: 3). Their simple and disturbing point is that our "free" uses of the internet and social media are valuable sources of data that internet service providers and social media platforms package and sell. Our online profiles, made up of expressed likes, dislikes, desires, and concerns, are highly marketable products.

Yet, as we will see, this inversion of the customer–provider relationship is only part of what is going on. Those who purchase the data we generate in the course of our online activity and our interactions with the internet of things are using that data in manufacturing and marketing new goods and services. But they are also using it to manufacture and distribute needs, longings, expectations, and ideologies. Systems of algorithmic agency are using our freely shared data to recursively craft our own adaptive conduct—in both online and offline environments—to better serve interests that are *not* our own. Bluntly assessed in terms of its power relations, the Attention

Economy 2.0 plausibly amounts to a system of digitally indentured servitude (Chisnall 2020).

At the very least, the Attention Economy 2.0 is conducive to the rise of what the legal scholar Bernard Harcourt calls the "expository society"—an organization of human community that is based on a progressive dissolution of the conventional boundaries among commerce, governing, and private life. In the expository society, our social energies are being systematically enticed and directed into "free" online spaces in which surveillance technologies that a generation ago would have been seen as coercive are "now woven into the very fabric of our pleasures and fantasies" (Harcourt 2015: 21). In a perfect instance of ironic self-colonization through the commodification of desire, the "technologies that end up facilitating our surveillance are the very technologies we crave" (ibid.: 228).

The expository society is an environment in which our lives are turned inside out. Our previously most private thoughts and fantasies are made globally public. Those who befriend and follow us most closely can be people we have never met. The things we seek out and decide to buy or learn have in fact already been chosen for us. Intelligence gathering practices that would once have been decried as violations of our rights to privacy are voluntarily accepted as normal. What we are giving away most freely is what costs us the most as our present increasingly becomes the future of a past we could never have dreamed was not of our own making.

Today, roughly two-thirds of the world population is connected to the internet and the average amount of time spent daily using social media, watching video streaming, listening to music and so on is 6.5 hours. According to the Nielson Total Audience Report for the first quarter of 2020, the average American devotes 12 hours daily to using the iconic devices of the Intelligence Revolution (smartphones, smartspeakers, computers, and tablets) and the no less iconic tools of more traditional electronic media (television and radio).[8] This is intelligence gathering at truly industrial scales. In Shoshanna Zuboff's (2019) acerbic wording, we have entered the "age of surveillance capitalism."

Often, this intelligence gathering is justified on the basis of streamlined access to goods and services combined with lower costs. And there is some truth to this. But much more is going on. All of our digital activity is being used, in real-time, to generate continually changing profiles of our patterns of attention. This intelligence gathering is not just good old-fashioned surveillance. The algorithmic processing of data about individual consumers in near real-time generates the commercial power of effectively "exteriorizing identity, separating it from the interiority of consciousness, and moving it into the realm of information machines" (Zwick and Knott 2009: 233).

Commercial consumer data services no longer provide their clients with broad-stroke population *profiles*; they produce flexible and responsive *simulations* of individual consumers for the purpose of precisely targeting and reflexively influencing their behavior.

Mass data collection is not new. Since the late nineteenth century, insurance companies have compiled data-rich actuarial tables for the purpose of identifying and classifying risk groups within various populations. The purpose of tabulating risk probabilities was, of course, to improve corporate bottom lines. But insurance companies were at the same time brokering the socially and morally laudable practice of distributing the costs of harms like automobile accidents and storm damage to homes and businesses, effectively diffusing responsibility for "bad luck." The algorithmic processing of consumer data today is amoral by design (Harcourt 2015: 147ff). Commercial intelligence gathering is not limited to gaining "epistemological power over consumers"; it also conducted instrumentally to acquire increasingly precise powers to "manufacture consumers ontologically" (Zwick and Knott 2009: 241). Data mining algorithms are the factories of the twenty-first century. They track consumers' attention and intelligence to differentiate among them with ever greater precision in order to predict and ultimately produce commercially beneficial differences in consumer desires and behavior.

It is these new powers that explain, for example, the massive shift of advertising expenditures from traditional broadcast and print media to the internet over the fifteen years. In 2001, for example, the total expenditure in the UK for online advertising was only 154 million pounds. By 2014, this had expanded to 7.1 billion pounds or 40 percent of the total advertising expenditures in the UK (Horten 2016: 25) That same year, the online advertising revenue for Google was greater than the total advertising revenue of all the newspapers in the world combined (Pasquale 2015). The power concentration potentials of the Attention Economy 2.0 are almost unimaginable. In 2017, of all new online advertising revenue in the world, 83 percent went to Google and Facebook, while 15 percent went to just three Chinese companies, Alibaba, Baidu, and Tencent. That left a meager 2 percent to be shared by all other online advertising platforms worldwide (Fischer 2017). Global online advertising revenues are now greater than those for television and far exceed the combined worldwide revenues of radio and print advertising.[9]

The commercial promise of precise and effective advertising is, of course, only one side of the equation. Algorithmic data mining is also transforming marketing and retailing. With state-of-the-art algorithms and predictive analytics, Amazon is able, for example, to set its price point for a given product below that of its competitors anywhere in the world. In effect, this means that local booksellers, music retailers, and clothing stores are forced to assume all the pricing risks of doing business while Amazon reaps the benefits of volume sales. But Amazon and other e-commerce platforms and online services like Orbitz are also able to capitalize on data about individual consumers by offering *higher* price deals to those they safely predict will be willing to pay. In short, companies like these are able to set revenue-maximizing price points for any given combination of consumer and product.

In addition, retailers like Amazon and Alibaba or streaming services like Netflix, iTunes, or Spotify are using machine intelligence to manufacture consumer interests and behaviors. Referring to these algorithmic systems as "recommendation engines" is disingenuous labeling. To take a single example, Netflix's algorithms are among the most sophisticated in the world and their operation is premised on the calculation that if a subscriber has not selected a film within 90 seconds, he or she will switch streaming platforms and the "recommendation engine" will have failed. The goal of the engine is not to make recommendations but to produce the target behavior of film selection and reduce subscriber churn to a predetermined limit. Netflix estimates that the effectiveness of its algorithm in reducing churn saves it over $1 billion annually (Harcourt 2015: 158).

Since 2013, YouTube, the other global leader in video streaming, has been using Google Brain—which carries out unsupervised machine learning—to run its recommendation engines. Changes initiated by the self-learning system resulted in successfully increasing the time individual users spent watching videos by 70 percent over the three-year period ending in 2016 (Grant 2017). Similar successes are being realized by music download and streaming services that make use of recommendation systems to identify and expand emerging fan bases, to influence musical tastes, and to accelerate consumption.

From the consumer/customer side of the equation, while the ability of leading Big Tech companies to lock-in dominant market share brings them immense commercial (and political) power, this is not generally seen as worrisome as long as this power is not used to artificially inflate prices. In fact, the price benefits to consumers on both goods and services is one of the reasons Big Tech has not yet been broken up through antitrust lawsuits (Wu 2018). But costs to consumers are not the only costs involved. As Jaron Lanier (2014) has noted, while purveyors and consumers of music content, for example, have benefitted hugely by recommendation-driven downloads and streaming, content creators have not fared nearly so well. In 2019, when digital streaming accounted for 80 percent of music revenues, Spotify—the current global leader in streaming subscriptions—paid content-providing musical artists a mere $0.00437 per play, less than half a cent. At the industry's highest per-play royalty rate of $0.01682 per play (by Pandora), an independent artist would need to be heard 87,515 times per month to earn the US monthly minimum wage or $7.25/hour or $1,472/month.[10] Structurally, the Attention Economy 2.0 is not conducive to guaranteeing that even creatively successful labor will be rewarded with a basic living wage.

Retail giants like Walmart have long used the epistemological and ontological powers of data mining to globalize and optimize their supply chains, driving supply costs down so far that individual suppliers are often only nominally rewarded for value added along the supply chain—a process, in effect, of manufacturing desirable suppliers. These same powers are now being used to manufacture desirable kinds of employees. Digital surveillance and algorithmic analysis are employed by Walmart to reduce "time theft" by breaking down all workplace duties into discrete, measurable tasks and subtasks and using time and motion studies to determine the fastest way of performing every subtask. Behaviors that do not accord with resulting "best practices"—like talking to coworkers or taking a roundabout way to the restroom—are grounds for discipline and dismissal. At Amazon "fulfillment centers," workers are outfitted with GPS-linked computers that specify routes to product shelves and set time targets for the subtasks of removing the item from the shelf, packing it, and placing it in the delivery chain. Workers are evaluated over spans as short as several minutes and those not measuring up to performance standards receive robot-generated texts to remind them that activities like speaking and catching their breath are not part of the job execution pathway (Head 2014).

At the other end of the labor spectrum, IBM's AI platform for professionals, Watson, is capable of conducting targeted research with superhuman efficacy. In one demonstration of its research power, Watson was provided with seventy thousand medical research papers about how specific classes of enzymes affect cancer growth and was tasked with predicting which enzymes would inhibit cancer growth. The papers Watson perused were all written prior to 2003, so that its predictions could be verified against actual scientific advances in cancer and enzyme research from 2003 to 2013. Based on its reading of these papers, Watson identified seven enzymes that it correctly predicted would inhibit cancer growth.[11] Seven years later, a machine learning system was able by itself to design a drug molecule for the treatment of obsessive compulsive disorder, which is now in clinical trials,[12] while a deep learning algorithm managed to discover an entirely new molecule effective in treating a wide range of antibiotic-resistant infections.[13]

The Second Machine Age: A New World or Business as Usual?

It's possible to argue that while these commercial applications of intelligence gathering may be new in scale and intensity, their impacts on the dynamics of competition within and across industrial sectors and on the future of work will be short term and not necessarily the revolutionary challenge to current economic, social, and political norms that techno-pessimists voice. Consider transportation. When wheeled transportation shifted from a reliance on muscle power to energies released from burning wood, coal, or petroleum products, transportation speeds changed dramatically. New types of vehicles were developed that took advantage of the new energy sources. Some jobs were lost: breeding and caring for draft animals, for instance. But others were created. The basic logic of transportation, one might claim, was not radically altered. It has remained about moving goods and people from point A to point B as quickly and efficiently/cheaply as possible.

By analogy, one might argue that although the commercial mining of consumer data that is now taking place at unprecedented scales with the help of machine learning algorithms, analyzing consumer behavior to maximize profit and competitive edge is as old as commercial practices. Walmart and Amazon's disciplining of workers is notable, perhaps for its greater precision, but it's actually nothing more than an update of Taylorist factory management practices that are over a century old, not some entirely new relationship between management and labor. Yes, intelligent technology will make some jobs obsolete. But it will create entirely new kinds of employment as well.

This way of looking at the impacts of historical changes in transportation technology and at the future impacts of intelligent technology errs in its exclusive focus on tools and their uses. A coal-fueled train may be just a mechanical version of a horse- or mule-drawn series of cargo or passenger wagons. A car may be just a "horseless carriage." But the transportation technology that came into being as tractors, trucks, and trains replaced vehicles pulled by draught animals also facilitated the co-emergence of new relational dynamics associated with, for example, the shipping-enabled transformation of agriculture into agribusiness, the growth of cities and then suburbs, the birth of cultures of mobility, and new familial norms. In short, these new vehicles were implicated in the construction of entirely new environments—physical and conceptual—as well as new social, economic, and political norms which have now shaped and been shaped in turn by the human experience for close to two centuries. The same is and will continue taking place with the growing ubiquity of the computational engines and industries of what MIT economists Eric Brynjolfsson and Andrew McAfee (2014a) have termed the Second Machine Age.

This might be a good thing. Although the Intelligence Revolution will produce significant economic and social disruption, the benefits of investing in it may still far outweigh the risks. In much the same way that the First Machine Age freed the human body from repetitive physical labor and drudgery, the Second Machine Age might free the human mind from repetitive mental labor and drudgery for other more exalted and gratifying pursuits. Our intelligent technology future is not merely one in which individuals and companies can survive. As Brynjolfsson and McAfee affirm in *Machine, Platform, Crowd: Harnessing Our Digital Future* (2017), it is one in which they can truly thrive. Properly "harnessed," AI might be used to bring about the end of want and support an epoch-making human renaissance.

This is a very appealing vision. Absent from it, however, are considerations of the motives and values of those who with one hand are feeding the general public's happy anticipation of a "smart" future's greater choice and convenience and who with the other are aggressively funding scientific and technical labors without which the Intelligence Revolution would remain fantastic fiction. What it leaves out—and cripplingly so in ethical terms—is the allure and pursuit of astronomical wealth and uncontestable power.

Commercial Empires: The Structural Bias of the Digital Network Infrastructure

In *Who Owns the Future?*, Jaron Lanier calls into question the structural equity of the digitally mediated attention economy, proclaiming that its network architecture is biased toward "privatizing benefit" while "socializing risk" (Lanier 2014: 279). It is an architecture that favors concentration, constrains real competition, and allows internet "sirens" like Google, Facebook, YouTube, Alibaba, and Tencent to get all of the benefits of a risk pool without any of its typically associated costs.

This bias of network structures toward monopoly is perhaps counterintuitive. But as Manuel Castells (1996) has noted, while the value of membership in a hierarchy is a function of distance from center or top of the hierarchy, the value of belonging to a network is a function of the quantity of nodes in the network and the quality of informational exchanges taking place through it. Thus, whereas the incentives for being part of a hierarchical organization *decrease* as the hierarchy grows, the opposite is true of networks. The incentives of belonging to a network *increase* as the network expands. This is what explains the mercurial ascent to market dominance of social media platforms like Facebook, Instagram. and WeChat.

A second feature of networks it that their growth or expansion is stimulated and shaped both by negative or stabilizing feedback and by positive feedback, which accelerates interactions and amplifies differentiation. In other words, network growth is stimulated and directed by the intelligence of its members—by their contributions to the network's adaptive conduct. When a network attracts a sufficient number and quality of members, its appeal reaches critical mass. Participation in it morphs from being an option to being accepted as a necessity, membership skyrockets, and competitors either are bought out or die the ignoble death of irreversible member attrition. In short, the structural default of network dynamics is an inexorable trend toward monopoly—a phenomenon now referred to simply as the "network effect" (Hindman 2018).

Thus, as the otherwise optimistic Brynjolfsson and McAfee are compelled to admit, the digitally mediated network infrastructure of the Attention Economy 2.0 is biased toward "winner takes all" competition through which great "bounty" is generated with minimal "spread" (Brynjolfsson and McAfee 2014a: 154). Bounty is systematically directed away from the providers of cheap labor and the owners of ordinary capital and toward those who either provide and manage network connections or who are fortunate enough to generate viral attention share with new products and services. As benefits are distributed in accord with the so-called power law, a vanishingly small number of winners in any given domain of business venture come to garner nearly all the rewards, followed by a "long tail" of anyone else attempting to stake a claim in that domain.[14]

Convincing empirical evidence for this account of the dynamics of the digitalized, global network economy can be found in the shift of global investment patterns over the course of the last two decades as the internet began generating commercial rewards. In 2001, the largest five corporations in the world, measured by market capitalization—that is, a company's stock market share price times total shares issued—were manufacturing, energy, banking, and retail giants: General Electric, Microsoft, Exxon, CITI Bank, and Walmart. By 2019, the largest seven corporations were all information technology and digital service providers: Apple, Alphabet (the parent company of Google), Microsoft, Amazon, Facebook, and the Chinese AI and internet enterprise companies Alibaba and Tencent.

Political Intelligence Interests

In a world where political fortunes depend profoundly on sustained economic growth, there is much more than money at stake. Just as the

Second Industrial Revolution destabilized global geopolitics sufficiently to create space for relative newcomers like the United States and Japan to gain leadership positions, the Fourth Industrial Revolution is being anticipated and embraced as an opportunity either to reengineer or commandingly reinforce existing global power structures.

Russian president Vladimir Putin succinctly summed up the stakes in a Knowledge Day talk to launch the school year in Fall 2017, stating that AI is the future of humankind and the nation that becomes the leader in it will rule the world. It is no accident that within a year China, a relative newcomer to the AI game, announced its intention to become a world leader in AI by 2030. China had been left behind in the Second Industrial Revolution and forced into the role of a pawn in the geopolitical Great Game being played among empires and nations in the final decades of the nineteenth century and the beginning of the twentieth.[15] It had been mired in the internal chaos of the Cultural Revolution when the Third Industrial Revolution in computing and telecommunications got underway. It is determined not to be left behind a third time (Tse 2017).

A New Great Game is underway. As it was during the competition among colonial and imperial powers in the late nineteenth and early twentieth centuries, the competition for supremacy in developing and deploying intelligent technology is a competition among distinct business-state-military alliances. But whereas the original Great Game was essentially a land grab with natural resources and labor as the primary objects of command and control, the New Great Game is essentially a time grab played for dominance in the attraction and exploitation of both data and attention—a struggle for power in what amounts to the colonization of consciousness itself.

Marriage Alliances: State and Corporate Intelligence Gathering

Given the history of national defense funding for AI, especially in the United States and the UK, it is not at all surprising that the revolutionary confluence among big data, machine learning, and AI is transforming state surveillance and security programs. While the commercial sector has been at the vanguard of exploring the new technological frontiers of intelligence gathering over the last decade, state intelligence gathering efforts have not been far behind.

Among the revelations in the Edward Snowden and Wikileaks files is the fact that—well before the "Cambrian" explosion in the evolution of learning algorithms—state surveillance systems in the United States and the UK were making extensive use of "deep packet inspection." That is, they were not only gathering metadata about who is connecting with whom digitally. With cooperation from wireless and internet service providers, they were also opening up the code packets used to transmit data and inspecting their content. Even a decade ago, the US National Security Agency (NSA) was annually intercepting billions of emails, phone calls, and text messages and was operating a data visualization system called the Treasure Map that was capable of mapping the entire internet, locating every internet-connected device in the world in real time (Harcourt 2015: 15).

Monitoring the content of internet traffic by means of packet inspection is not necessarily a nefarious act. It is standard practice among the providers of both internet service and antivirus and spyware protection. What is noteworthy about state surveillance uses of packet inspection is that government agencies effectively depend on the willing cooperation of private sector companies in order to gain access to digital communications traffic. This marks the emergence of a new mode of governmentality that intimately merges the commercial interests of the communications and security industries with the surveillance interests of nation-state agencies—interests that are both epistemological and ontological.

State surveillance is generally practiced for the dual purposes of cross-border and within-border security. In the former case, intelligence is gathered as a means to and as an expression of what we have referred to as epistemological power. In the United States, for example, the NSA makes use of all available technology to monitor, collect, and process data generated by state and nonstate actors whose actions are deemed to be of strategic or security relevance. The NSA also monitors data traffic within American borders for the same reason. Most countries that can afford to invest in smart surveillance—as opposed to labor intensive surveillance carried out by humans—are doing so to improve national security. And while nations vary in terms of the extent to which they can legally gather data about their own citizens, when it comes to gathering and analyzing data about noncitizens, most states are generally less concerned about what *should* be done than they are about maximizing how much *can* be done. This includes, as we will see, making ontological uses of systems for cross-border digital surveillance to infiltrate and either shape or destabilize informational environments in other countries.

Within-border state surveillance can also be conducted to gain and wield epistemological and ontological power. Perhaps the most open and aggressive instance of this in US history occurred during the virulent anti-communist campaign that was led by Senator Joseph McCarthy from the late 1940s through the mid-1950s, combining old-fashioned surveillance and existential scare tactics to discipline the thinking of the American public. While the United States and other liberal democratic governments are not apparently making similar ideological uses of smart surveillance to discipline public thinking, that is not true in illiberal states like Russia, Iran, and China, where information circulations are subject to increasingly sophisticated, algorithmic censorship. The ontological powers of digital connectivity are being avidly wielded in any number of autocratic regimes (see, e.g., Kendall-Taylor, Frantz and Wright 2020).

To date, the most explicit and proactive use of smart surveillance for ontological purposes has been by the Chinese government. Since its inception, the Chinese communist party-state has regarded gathering data about individual citizens as its unquestionable right and controlling circulations of information has been a mainstay of Chinese political culture—a mechanism for reining in conduct inconsistent with the party-state's interests. With the exponentially growing data resources resulting from the Chinese people's enthusiastic embrace of market competition and digital connectivity, and with the new analytic and predictive powers afforded by machine learning and AI, the Chinese state is shifting from a culture of informational constraint to one of smart statecraft aimed at achieving dynamically engineered and yet stable societal well-being. Partnering with private corporations like Alibaba and Tencent, the Chinese government is developing and testing prototypes for an algorithm-driven "social credit system" that have the win-win purpose of producing both more "trustworthy" citizens and more reliable consumers. In fact, similar state–business partnerships are explicitly crucial to China's national AI strategy of developing innovation ecologies in which intelligent technology can evolve in alignment with the state's interests.[16]

This blending of commercial and state interests is not an illiberal anomaly. Autocracies are not alone in cultivating corporate-state alliances that use smart surveillance and digital connectivity to exercise ontological power for political purposes. There is now no disputing the fact that American and other democratic elections have come to be battlefields on which competing political interests wage AI and machine learning campaigns. While the extent and effectiveness of the work done by Russian botnets to shape American public opinion and voting behavior during the 2016 presidential election campaign is open to debate, there is no debating the seriousness of domestic investments in digital politics. During that campaign, 33 percent of pro-Trump Twitter posts were botnet generated and 22 percent of Clinton tweets had similar robotic origins. Cambridge Analytica, a company that (according to its website) "uses data to change audience behavior" and that registered its first major political success in algorithmically swinging the Brexit vote in the UK, used Facebook member data in the run up to the 2016 US presidential election to compile detailed personality files on some two hundred twenty million Americans and target them with digital news feeds intended to affect their voting behavior. In a telling illustration of the marriage of commercial and state interests, Robert Mercer, the major donor to the Trump campaign, was also one of Cambridge Analytica's founding shareholders, while Steve Bannon, Trump's one-time senior counselor, sat on the company's board of directors (Hussain 2017). The digital marriage of commercial and state interests does not bode well for democracies founded on the viability of truly free and fair elections.

In an incisive study of how the Intelligence Revolution is impacting American society, *The Black Box Society*, legal scholar Frank Pasquale (2015) documented the data-driven perforation and gradual erosion of the boundary between state and commerce along three key dimensions: reputation, search, and finance. In it, he offers a lucid exposition of the liabilities of infusing society with "black box" computational systems whose workings are too complex or opaque to understand, including the ways in which algorithmically generated credit scores and social reputations can perpetuate historical injustices and inequalities and the ways in which proprietary algorithms in the finance sector can result in regulatory obfuscation.

He also details how legal restrictions on state surveillance in the United States are being circumvented by the creation of "fusion centers" where, in the name of information sharing, government agencies marry their constitutionally limited data with the unregulated collections generated by private industry (Pasquale 2015: 46). Among the spinoffs of such fusion centers are private companies devoted to "threat" identification. To take a single example, the company Recorded Future was launched with a combination of commercial funding from Google and state funding from the Central Intelligence Agency (CIA) and maintains a database of billions of continually updated and indexed facts (Shachtman 2010). This data is analyzed by its proprietary Threat Intelligence Machine, which uses machine learning and natural language processing to identify threats from anywhere in the world in real time.

Pasquale's well-supported conclusions are that leading "Internet and finance firms present a formidable threat to important values of privacy, dignity, and fairness" and that "[t]his threat, now increasingly intertwined with the power of the government, is too often obscured by self-protective black box practices" (Pasquale 2015: 215). He traces this threat in the United States to regulatory failures caused in significant degree by circulatory flows of people among government offices (both appointed and elected), the corporate lobbying industry, and business and finance circles. In his colorful characterization, the traditional tug-of-war between corporate and government interests has become a *pas de deux*: a seductive "dance for two" in which the same people can, over the span of several years, switch from one side to the other (Pasquale 2015: 215). In the climate of emergency responses to the COVID-19 pandemic, there is considerable potential for these circulations of people to become "hardwired" as part of the circuitry of government services.

Although there are significant differences from nation to nation and across regions, the relationship between commercial and state surveillance is surprisingly consistent. The simple reason for this is that when governments do not own and operate the digital connectivity infrastructure, if they have any intention of using data flows—whether for surveillance or public administration purposes—they have no recourse but to rely on commercial wireless, internet, and social media platform providers who thus necessarily acquire positions of considerable structural power. At the same time, however, the right to gather and monetize personal data—which is crucial to commercial success in the information and attention economy—depends on securing appropriately lenient government stances on privacy and personal data property rights.

This is true even in the European Union with its strong commitments to liberal democratic politics and social welfare systems; bargains generally have been cut between government and industry in ways that allow both state and corporate exploitations of data to proceed. Thus, although the EU's General Data Protection Regulation, which passed into law in 2018, is quite explicit in its aims of transferring at least nominal data ownership rights from connectivity and platform providers to individual citizens, the EU High-Level Expert Group on Artificial Intelligence has developed a comprehensive "European Strategy on Artificial Intelligence" that seeks to balance desires for ethical, legal, and societal assurances regarding the human-centered development of AI with perceived political and commercial imperatives for developing a distinctively European "brand" of AI that is capable of cornering a commanding share of the global AI market. The "dance for two" can be done to European as well as American "accompaniment."

Smart Government: The Appeal of Algorithmic Public Administration

The United States, China, and the EU offer very different visions of the marriage of corporate and state interests and of how intelligent technology can and should be deployed to best serve their defining national, regional, and global political ambitions. A primary vector of difference is the extent to which the state assumes responsibility for determining the values that shape the evolutionary dynamics of intelligent technology and the extent to which the state is content to allow market and civil society forces to shape those values. Generally shared among these different visions, however, is the acceptance that accelerating the development of intelligent technology is imperative politically for state security and global status purposes and that integrating intelligent technology into government services and program administration is imperative both functionally and fiscally as societal dynamics become both more complex and costly to manage.

Governments worldwide are turning to big data and AI to solve public administration problems, opening pathways to futures in which government on behalf of the people might not be conducted primarily by people. Administration by algorithm can save time and money. Smart welfare and public assistance programs taking full advantage of data analytics can streamline service delivery and reduce the impacts of human error and prejudice. Algorithmic watchdogs can greatly expand the scale and scope at which government regulation enforcement is conducted. Yet, the same systems of intelligence gathering that allow these benefits to be sought also can be used to engage in digital dictatorship and, if not extremely well-designed and subject to continuous oversight, have potentials for perpetuating precisely the kinds of inequalities and insecurities that many public administration programs are intended to address (Eubanks 2019).

In keeping with the New Great Game narrative and encouraged by successes in the marriage of intelligent technology and government thus far, the boldest advocates of industry-state accommodation go so far as to envision the emergence of a new global order comparable to the "peace" secured by the informational networks of ancient Rome and imperial Britain—a "Pax Technica" founded on "a political, economic and cultural arrangement of society's institutions and networked devices in which government and industry are tightly bound in mutual defense pacts, design collaborations, standards setting, and data mining" (Howard 2015: 145). Those more skeptical about this marriage point to the dangers of concentrating powers of force, scrutiny, and perception control behind algorithmic veils, suggesting that if the great political debates of the twentieth century turned on "how much of our collective life should be determined by the state and what should be left to market forces and civil society," the great debates now should center on "how much of our collective life should be directed and controlled by the powerful digital systems and on what terms" (Susskind 2018: 361).

Smart Citizenries and the Absence of Political Resistance

At least in liberal states, the existence of complex and often compromising relations among government, industry, and the public of the kind exposed by Pasquale might be expected to result in grassroots political resistance. Yet, for the most part, network and social media platform consumers have not engaged in mass protests or in concerted lobbying, for instance, on behalf of stricter regulation of personal data use, whether by government or industry. On the contrary, internet and social media users globally have proven to be largely content to tolerate default invasions of their privacy in exchange for the ease and effectiveness of online experiences that have become not merely habitual but absolutely crucial to the formation and maintenance of their own social and personal identities.

While the causes for this complacency are undoubtedly quite complex, it is instructive to consider the historical case of prereunification East Germany, where empirical studies conducted by the government showed that access to foreign media led to a net increase of support for the regime, rather than calls for its reform or removal (Kern and Hainmueller 2009). Indeed, it can be argued that the shifts from mass media like network television and radio, to cable television, to streaming services and other forms of web-delivered entertainment have had the practical effect of allowing consumers/users to progressively "opt out" of politics (Prior 2007).

These claims run directly counter to the widely embraced belief—most passionately expressed during the early days of the so-called Arab Spring—that social media can play important roles in galvanizing dissent and opening spaces of citizen group empowerment. In fact, the emancipatory political role attributed to social media seems to have been more a matter of modern myth-making than a reflection of on-the-ground political realities (Morozov 2012). The weight of evidence globally is that political claims lodged on the internet by citizen groups have been modest (if not trivial) in scope and have done little to pressure political and corporate leaders into enacting real social reforms. Indeed, they seem instead to have had the ironic effect of increasing "the hegemonic authority of bureaucratic actors who recognize that the Internet creates a diversion that minimizes the likelihood of street-level response" (Maratea 2014: 124). As Evgeny Morozov pointedly notes, "The most effective system of Internet control is not the one that has the most sophisticated and draconian system of censorship, but the one that has no need for censorship whatsoever" (Morozov 2012: 58).

None of this is to deny the potential of global networks and social media to generate awareness of failures and threats to social justice or to organize collective action. These potentials have been carefully examined by Manuel Castells in *Networks of Outrage and Hope*, and they are substantial. Yet, as Castells (2015) points out, the work of social justice requires the exercise of social autonomy in actual, and not only virtual, public spaces. To be both effective and sustained, commitments to social justice must be socially embodied, not only digitally expressed. The reality today, however, is that even when urban spaces are claimed for a time in open resistance to either the corporate or the state partners in the dance of exercising epistemological and ontological power, they are dwarfed by the "spaces of action" that are being created by the intelligence industries in support of both the attention economy and the surveillance state.

Resistance may not be futile. But for reasons we will explore in the next chapters, a profound change may be required in who we engage in resistance as. We will, at the very least, need to avoid sequestration within compulsively desirable and attention-retaining "filter bubbles" (Pariser, 2012) to be able to engage in shared and sustained critical deliberations of the kinds and depths needed for global predicament resolution. That is no small matter. But it will also depend on how far and how fast current efforts go in weaponizing machine learning and AI. There is, in fact, a third partner in the dance of negotiating intelligence advantages.

Military Intelligence Interests

While the Cold War "Iron Curtain" hung between competing geopolitical hegemons, with weapons of mass destruction conspicuously in sight, the "Digital Curtains" that separate competing corporate/state alliances in the New Great Game have been lowered gently and ubiquitously in our midst, their "weapons of math destruction" (O'Neil 2016) nowhere to be seen. This is not to say, however, that in these alliances there is no one dreaming of artificial soldiers as well as savants and seers, contemplating the uses of AI weapons of mass destruction and disruption.

Long before the intelligence-gathering dance of commercial and state actors began in earnest, AI research and development was being funded with military applications in mind. This continues to the present day. Direct US Department of Defense spending on unclassified AI research is running roughly $2 billion annually. Research in robotics, cyberwarfare, new generation drones, and autonomous weapons development is being funded at similar levels, while something on the order of $4 billion is being spent on classified AI-related projects. Yet, while the funding to the Pentagon and to DARPA is substantial, as is noted in a 2017 report commissioned by the Intelligence Advanced Research Projects Agency (IARPA), private sector spending on commercial applications far exceeds that of government spending, and leveraging this is understood as crucial to securing American military advantage. Military technology, the report stresses, is going through a revolution as great as those brought about previously by airplanes, nuclear weapons, and biotech.[17]

Other states, in both Europe and Asia, are also investing in military applications of AI. But by a considerable margin, the other major state actor in military AI applications is China. Over the coming years, the Chinese government's direct spending on military applications of AI is estimated to be roughly equivalent to American spending levels. Within the People's Liberation Army (PLA), the conviction is that a "singularity" will occur on the battlefields of the future—a point at which humans will not be able to keep up with the pace of decision-making by autonomous weapons and artificial strategy agents. The PLA is committed to making sure that China will be able to assert power parity, if not a decisive advantage, when that time comes (Kania, 2017).

In keeping with the IARPA report in the United States, the Chinese government has been cultivating and leveraging targeted military-civil fusions aimed at taking full advantage of commercially developed advances in machine learning and AI to "leapfrog" into a position of intelligent military dominance. China's approach to achieving military-civil fusion is distinctive, however, in benefitting from a steady current of Chinese AI and machine learning experts flowing through both commercial and military research and development programs in the United States and elsewhere, and then back into China (Kania, 2017). This circulation affords China worldwide footholds at the cutting edges of new research and development.

Globally, the dynamics of civil-military fusion will be crucial to determining the nature and scope of what is generally regarded as an inevitable AI "arms race." Although direct military spending on robotics, autonomous systems, and cyber security can be expected to grow in the coming decades, this spending will remain modest in comparison to relevant research and development funded by private sector commercial interests. For example, global research and development spending on the aerospace and defense sector in 2016 was approximately $30 billion. This is the sector most directly associated with the design and development of weapons systems, military aircraft, and so on. By comparison, global research and development spending in the information and communications sector was just over $200 billion, and spending in the automotive sector was just under $100 billion (Cummings 2017).

What this means is that the core systems of smart intelligence-gathering, decision-making, and navigation needed to develop fully autonomous weapons systems—for example, drones, fighter-bombers, tanks, or landing assault vehicles—are being developed in commercial, rather than military, research settings. Facebook and Amazon both have drone development research projects. Uber, Google, and Apple are all involved in autonomous vehicle development. To supplement existing in-house expertise, these and other major corporations are buying talent—and, at times, entire enterprises—out from under direct military research contracts. Google, for instance, recently purchased Boston Dynamics, a company long funded by the US military to do cutting edge work on robotics (Cummings 2017.) A major implication of this disproportionate prominence of commercial interests in the domains of machine learning, autonomous systems, and cyber security is that specifically military applications will be assembled out of what amount to "off the shelf" components manufactured at scales and with quality controls commensurate with realizing reasonable return on investment. In short, the basic components for building military-use autonomous systems may be within the reach, not only of nation-states but also effectively unrestricted numbers of nonstate actors.

The classified nature of core dimensions of military research and development makes it difficult to assess the degree to which military dreams of smart weapons and human–machine symbiosis are being realized. The US military has tested drones that are able to fly into urban environments and differentiate among mission-identified targets, noncombatant civilians, and other clear mission threats—discerning, for instance, the difference between a person aiming a camera at the drone and someone aiming a pistol at it. They do this without any human intervention. Robotic fighter jets and missiles that decide on what to destroy have been tested. AI pilots running on cheap handheld computers have been able to defeat US-trained fighter pilots in combat simulations. And swarm intelligence systems have been developed that enable autonomous drones—which can be large enough to carry significant explosives or as small as bees armed with poisonous darts—to approach targets individually or as a squadron that disperses if detected, with drones proceeding independently to their target(s) and reassembling as needed to accomplish their mission.[18]

While the capabilities of autonomous weapons systems like these are at present relatively modest and prone to errors of judgment, they offer proof of concept: it is possible to build lethal, self-navigating devices that can identity targets and take aggressive action entirely on their own (Rosenberg and Markoff 2016). What they point toward is the goal of repositioning humans with respect to the so-called OODA loop: Observe, Orient, Decide, and Act. The ultimate dream is an artificially intelligent system that will be able to engage in real-time planning, adapt to emerging mission realities, and carry out objectives at matchless speed and with inhuman determination.

Intelligent weapons systems are already ushering in a new era of battlefield dynamics. The United States, for example, is already able to deploy a close-in weapons system that, once engaged with human input, will track and fire autonomously at incoming conventional targets. But once incoming targets are also "smart," fully autonomous defense systems will become tactical necessities since the "smart" evasive actions of incoming enemy aircraft or missiles will be impossible for humans to track. The rates of human information processing, decision-making, and decision-execution are simply no match for AI. This is what accounts, for example, for the effectiveness of automated high-speed trading on stock and securities markets. Intelligence operating on an electronic substrate is orders of magnitude faster than intelligence operating on a biological substrate.

It can be expected that the progressive use of machine learning and AI on the battlefield eventually will have the effect of so powerfully compressing the OODA cycle in the direction of "instantaneous situational awareness" that humans will be forced completely out of mission design and implementation (Hussain 2017: 99). Taken to its logical extreme, the likely result is that, in the military domain, as in others, the role of humans may be reduced to that of merely explaining what has been done and why.

In contexts where militaries are subject to political oversight and public scrutiny—as is at least nominally true in the United States—strategic and tactical horizons tend to be coincident with their explanatory horizons. That is, ontological power tends to be constrained by epistemological power. In anticipation of political blowback from deploying weapons systems built around algorithmic black boxes, DARPA has recently launched an initiative to develop what it calls Explainable Artificial Intelligence (XAI), and the Pentagon has been vocally committed to keeping humans in the OODA loop. As Deputy Defense Secretary Robert O. Work has stated, the ideal at this point is to be able to combine the "tactical ingenuity of the computer to improve the strategic ingenuity of the human"—creating a "Centaur" weapons system in which a human soldier is synched with a robotic body or fitted with an exoskeleton to be able to deal with battlefield unpredictability from within continuously updated, intelligence-saturated augmented realities, implementing human directives with machine-amplified force at lightning speed (Rosenberg and Markoff, 2016).

If this all sounds like something out of a science fiction film, that is appropriate. Within the Pentagon, the militarization of AI is regarded both as a national strategic necessity and as an ethical and legal dilemma—one that Pentagon insiders have dubbed the "Terminator conundrum." The general consensus is that, at current rates of advance, a fully autonomous "killer robot" will not be technically feasible in less than eight to ten years. The conundrum is how to proceed in the meantime. Should intelligence-weaponizing research and development be allowed to proceed at its current pace, driven primarily by commercial interests? Should this research and development proceed more aggressively with significant increases in targeted research funding? Or should the appeals of many scientists and ethicists be heeded to either slow or completely curtail the militarization of autonomous systems?

These are all valid questions to which different stakeholders are likely to offer very different responses. The geopolitical conundrum is that affirmative answers to either the first or third of these questions could be disastrous if other state actors are aggressively accelerating the militarization of AI. Can "we" afford to lose competitive edge when the stakes are likely to be "winner takes all"? This is the sleep-depriving question that is being asked in military circles, not just in the United States but in China, Russia, Europe, the UK, Israel, Saudi Arabia, and Iran. The emerging consensus in military and security circles at present seems to be that "it is only AI that can protect us from AI" (Hussain 2017: 156).

Given that the core components of intelligent weapons systems will be built commercially and made available at relatively low cost, however, it will be entirely within the capabilities of nonstate terrorist groups to develop and deploy swarm weapons, for instance. Given also the winner-takes-all stakes of achieved superiority in autonomous weapons development, even if tactical robotics development is cost-prohibitive, it is a virtual certainty that "bad actors" will develop maximal capabilities in autonomous weapons and cyber-warfare. The "rational" conclusion, then, is that an AI "arms race" is unavoidable. Bans and treaties are unenforceable and thus accelerated research and development are imperative.

This conclusion is disturbing, to say the least. It recalls the Cold War "solution" of nuclear weapons buildup on the theory of first strike deterrence by threat of mutually assured destruction. Only in this case, the commercial availability of much of the basic components of intelligent weapons places the bar to entry very, very low. Intelligent weapons *technology* will bring about conditions of readiness to employ, and not merely deploy them. No less disturbing, however, is the admission within military circles that building intelligent ordinance-delivering tools ultimately will only yield linear growth advantages. Smart missiles are still just ordinance-delivery mechanisms for destroying property and people. But co-opting the commercial infrastructure of digital connectivity and using it run global-scale influence engines opens prospects for directly affecting public opinion through precision-targeted misinformation, and this is capable of yielding exponential advantages by directly undermining the cognitive integrity and political will of competitors. Building on commercial lessons in profitably manipulating attention, feeling, and thinking at lightning speed with algorithmic content generation, personalized targeting, and firehose dissemination, a new kind of arms race is already underway: a race in which mental munitions are fired by automated weapons of mass influence in a theater-less, dissent-inducing global assault on shared experience (Telley 2018).

Whether this so-called realpolitik approach to dealing with the security predicaments posed by autonomous systems is truly rational is debatable. For now, it is enough to register that it is a purely technical approach in which legal and ethical considerations are set aside under the all-forgiving umbrella of "special circumstances." If we look back to other arms races over the course of human history, the lesson is that technology wins. The arms in question might have been used at industrial scales, wreaking Armageddon-like destruction, as was the case with aircraft. Or such mass deployment might have been avoided, as has thus far been the case with biotech weapons. But all of the technologies within which previous arms races have been run have persisted.

The Theater of Competition in the New Great Game: Consciousness

Great differences exist around the world in the three-way dance among commercial, political, and military interests in intelligent technology. The intelligence-infused environments of digital connectivity and opportunity are perhaps as distinct as national and regional cuisines. In fact, a compelling case can be made that there is not, in actuality, a single internet but rather a number of competing, ostensibly worldwide webs of digital connectivity and governance, each of which supports different kinds and scopes of digitally mediated agency in keeping with a legitimizing narrative of future benefits (O'Hara and Hall 2018). At present, the Intelligence Revolution does not seem likely to result in a global Pax Technica, but rather in deepening competition among, at the very least, actors committed to constellations of values variously associated with the connectivity regimes of Pax Americana, Pax Sinica, Pax Slavica, and Pax Europa. In each, distinctive dances will take place among commercial, political, and military elites and interests. In each, the music to which these partners will dance will be amplified by industrial flows of attention energy and contoured by the data carried along with them.

These differences of approach to smart government and smart capitalism are important for us as citizens and consumers—differently positioned as we are in terms of national residence, class, gender, and culture. But they should not blind us to the structural dynamics shared by these various trajectories into increasingly smart futures and the risks associated with them. Regardless of their ideological colorings, corporate, state, and military interests in intelligent technology around the world have in common fundamental commitments to the values of predictive precision and behavioral control. Different narratives may be deployed to ensure and accelerate citizen and consumer buy-in but data mining and the progressive transfer of social, economic, political, and cultural energy from material to digital environments are embraced by all.

To better appreciate what is at stake for humanity in the New Great Game, regardless of who "wins," it is useful to return to the contrast made in passing earlier between the "land grab" or spatial politics that characterized the Great Game that began one hundred fifty years ago and the "time grab" or temporal politics that characterizes the New Great Game. In the former, competition was visibly manifest. It involved the movement of goods and people across land and sea, and what was at stake were spheres of physical influence. This was truly *geo*political competition that eventually resulted in the global depredations of two very "hot" World Wars, followed by an extended series of Cold War proxy conflicts. In the New Great Game, competition is essentially invisible. It involves computational movements of data and the mass export of attention into sensory environments that are not substantively physical. Indeed, they are not essentially spatial.[19] Granted sufficient bandwidth, in contradiction of one of the basic laws of physical presence, there are no limits to how many people can visit a given website or social media platform at any given time. There are no limits to how many films or pieces of music can be within equidistant digital reach through digital streaming services. That is, in contrast with physical space, where two things cannot occupy the same space at the same time, network connectivity permits infinite superposition, and this effectively permits what amounts to instantaneous "travel" from anyplace in the net to any other place. In fact, the only "distances" evident in so-called cyber*space* are temporal—they are not measures of how far one must go to get to a desired destination, they are measures of how long it takes to form and express an intention to get there. Digital environments are essentially temporal environments. The New Great Game is *chrono*political competition—a competition for control of the textures of lived time.

Crucially, digital environments are also wholly intentional. All digital environments and all of the sensory connections possible in them have been purposely coded into existence. Even if a randomization program is built into a given environment's dynamics, that randomization program was itself purposely encoded into the environment's computational structure. While a physical environment can be wholly controlled only if it is completely closed, there is a sense in which digital environments are both hyper-open and meta-controlled. The nonspatial character of digital environments makes possible not only the instantaneous delivery on our connective choices but also the intentional and profitable elimination of human effort and all of the physical and social frictions involved in it. Ironically, the more freely we connect digitally, the more fully we are subject to invisible and unfelt algorithmic steering.

Taken together, these peculiarities of digital environments suggest that however innocently it might have begun, the mass transfer of social energy from physical to digital forms of engagement now amounts to a mass expatriation of social agents to environments in which their every move and intention can be surveilled, remembered, and potentially manipulated. Digital connectivity is not merely a "time sink," it is a medium for time theft. This aspect of the Attention Economy 2.0 was described earlier in terms of the lost investment potential of the time and attention capital that digital media effectively export out of our homes and neighborhoods. In light of the fact that our time and attention are not coerced out of us, but given freely, it might be objected that this loss of capital is not really time theft, but rather a market exchange of certain sets of temporal possibilities for others. Some may benefit more than others from such time-and-attention market transactions, but benefits are quite obviously enjoyed by all involved. Nevertheless, given the thoroughly intentional and profitably structured nature of the digital environments to which we are offered access in exchange for our time and attention, and given the Buddhist understanding of the relational nature of all things, including consciousness, it is arguable that what appears to be a fair and legal process of market exchange is more accurately seen as a process of colonial expropriation.

The nature of consciousness and its causal conditions are topics of extensive contemporary scientific and philosophical debate. In keeping with contemporary arguments on behalf of seeing consciousness (or mind) as irreducibly embodied, enacted, and environmentally embedded, Buddhism is generally committed to regarding consciousness as a function of the differentiation of among sensing presences and sensed environments.[20] Consciousness is, in other words, neither singular nor something internal to a sentient being. It is a qualitatively dynamic *mode of relationality*.

The basic modes of human consciousness are visual, auditory, tactile (touch), olfactory (smell), and gustatory (taste). Visual consciousness, for example, arises with the dynamic interplay of a visual organ (my eye) and a visual object (the Pacific Ocean horizon beyond the buildings of Honolulu)—the realization of dynamic visual relations. These five sensory consciousnesses are not distinct "spaces" in which different kinds of sensory events happen to take place; they are qualitatively distinct modes of happening, distinct temporalities. Cognitive consciousness, according to Buddhism, is our sixth sensory modality—one that emerges with the interplay among thinking/discriminating presences and the environments constituted by the dynamics of the other five sense modalities. To these six consciousnesses, the Yogācāra school of Buddhism adds two further modes of relationality: self-consciousness as a function of sensitivity to changes in the dynamics of cognitive consciousness as an environment of experiential possibilities, both desirable and undesirable, and what might be called the karmic consciousness, which emerges as the dynamic interplay of self-conscious presence and the essentially temporal environment of outcomes and opportunities made evident through sentiently enacted values and intentions.[21]

One implication of this relational conception of consciousness is that the evolution of human consciousness is not best understood as taking place objectively in a preexisting world but rather as a dynamically improvised process of exploring sensory wilderness; of extending sentience beyond familiarly enacted visual, auditory, cognitive, and other kinds of sensory relations; eliciting environmental responses to our exploratory efforts; discerning qualitatively among them; and then furthering our efforts. Digital environments are likely soon to support all six basic sense modalities. They are already environments within which we are enacting explorations of the relational dynamics of social, political, economic, and cultural engagement. But, as computational constructs, none of the environments into which we are being actively solicited are truly wild. They are environments intentionally opened and now increasingly attuned algorithmically to potentials for maximizing attention capture and capitalizing on the control that this affords for shaping human temporality.

If colonization is the appropriation of a place or a domain and all that it includes for one's own use and purposes, then there is a nonmetaphorical sense in which what is ultimately at stake in the New Great Game is the colonization of consciousness—an appropriation of the domains of human attention and sensory experience by artificial and synthetic intelligences acting on behalf of human (commercial, political, and military) interests that has the potential to transform us from the inside out. Ultimately, the colonization of consciousness is a colonization of interdependence itself: a colonization of the relational totality of which our physically and socially embodied presences are constitutive parts.

4

Total Attention Capture and Control: A Future to Avoid

The course of the Intelligent Revolution is not predetermined. The karmic nature of the human-technology-world relationship ensures that our technological direction is not fixed. We do not have a manifest technological destiny. The karmic nature of this relationship gives us assurance as well that changing its course will involve enacting different values and intentions than we have been.

Dreams of artificial intelligence (AI) began as and largely continue to be dreams of wish-fulfilling agency. Over much of the two millennia following Homer's imaginings of machine labor and intelligence, muscle power could be augmented or leveraged but not entirely replaced. It was only with the invention of steam and internal combustion engines that it became possible to begin replacing muscles with machines at mass scale and substituting fossil energy sources for food. Humanity's capabilities for wealth generation expanded almost miraculously. Global standards of living rose dramatically, decades were added to average life expectancies, and attention energy that had been devoted to meeting basic subsistence needs for most of human history were able to be channeled into the arts and music.

The scientific advances that made the machine revolution possible also naturalized the order of the cosmos, making it seem feasible to build machines that leveraged the laws of thought as well as they did the laws of nature—machines able to do mental, and not just brute physical, labor. Once algorithms could be run on electronic substrates rather than mechanical ones, computations could be performed at speeds and with accuracies far exceeding those achievable by humans. Finally, with the data resources needed to realize the potential of deep machine learning, it became possible to build electronic savants that could learn on their own and even acquire the predictive talents of seers.

Pair these machine intelligences with technical advances in three-dimensional printing, bioengineering, and neuroscience and it not hard to envision a near future in which human creativity accelerates to the point that we finally reach escape velocity—not from earthly gravity but from the psychic or spiritual gravity of practical necessity. Machines are now being built that can discern what we are feeling nearly as well as other humans do, reading our attraction and distraction, our pleasure and displeasure, and translating our intentions into real-world actions. Moving forward on this course, it does not seem implausible that the dream of ambient and ever-ready wish-fulfilling artificial agencies might finally come true.

There is much that is appealing in this triumphal vision of humanity freed from effort. In what follows, I want to caution against it, making use of several fictional narratives to sharply imagine in some detail the future(s) into which we might be carried if we permit the Intelligence Revolution to remain on its current course. There are, to begin with, legitimate practical concerns that this course seemingly involves accepting the winner-takes-all inequalities of the networked attention economy, the risk of mass unemployment and underemployment, and the cold technical rationality which would insist that we let AI "take the lead in making decisions ... and let people take the lead if others need to be convinced or persuaded to go along with these decisions" (Brynjolfsson and McAfee 2017: 123). But, in addition, there are two more philosophical concerns. One is the karmic concern that wish-fulfilling machines will give us *exactly* and *only* what we wish for. The other, perhaps even deeper concern is that remaining on course for an increasingly "smart" future of effortlessly satisfying lives will result in the atrophy of intelligent human practices, including those for ethically creative predicament resolution. We may forfeit our capacities for course correction.

The Predicament of Wish-Fulfillment: The Midas Touch and Errant AI

In addition to the story of Pygmalion, Ovid's *Metamorphoses* includes the myth of King Midas, to whom we owe the idiom of having a "golden touch." Midas was renowned as a lover of fine living whose rose garden was deemed the most beautiful in the known world. According to legend, a satyr by the name of Silenus was discovered passed out in Midas's garden after a night of carousing. Learning that Silenus had been the teacher of Dionysius, the god of wine-making and fertility, Midas delightedly asked him to join in a few days of merrymaking, after which he escorted Silenus back to Dionysius. Immensely grateful to have his teacher returned in good health and spirits, Dionysius offered to fulfill any wish Midas cared to make. Midas's famously greedy wish was to have everything that he touched turn to gold. Upon returning home, he discovered that while he was able to turn any twig or flower he touched into a masterpiece of the goldsmith's art, the same thing happened to the food he tried to eat and the wine he tried to drink. Horrified, he returned to Dionysius to ask that the wish be undone. The simple moral of the myth, of course, is to be careful what you wish for because you very well might get it. This is especially true if, unlike Midas, you are unable to have your wishes revoked.

Nick Bostrom (2014), in his book *Superintelligence: Paths, Dangers, and Strategies*, discusses a number of different scenarios for what might happen if and when we create an artificial superintelligence. He entertains utopian scenarios in which a superintelligence solves all of humanity's resource needs or becomes a benevolent dictator bringing peace and wellbeing to all, as well as dystopian scenarios in which a malevolent superintelligence remakes the planet to suit its own interests and either places human beings in zoos or eradicates them with the same kind of impunity with which we exterminate ants and roaches in our homes. Among the more far-fetched scenarios is one in which humanity is accidentally eliminated because an artificial superintelligence brilliantly pursues its humanly programmed goal of making paperclips. Like an artificial "Midas," everything this superintelligence encounters is turned into paperclips, first the entire planet Earth, then the rest of the solar system, the rest of the galaxy, and so on ad infinitum.

The philosophical point of the paperclip scenario is that granting even apparently innocuous wishes might have profoundly troubling consequences. We do not need to be as greedy as Midas for our wishes to backfire. Consider the quite simple wish to be freed from having to remember people's phone numbers. One way to accomplish this is the old-fashioned practice of keeping an up-to-date, hand-written ledger of phone numbers for friends, family members, colleagues, and so on. Those of us who have used personal phonebooks know that, over time, frequently dialed numbers become imprinted in muscle memory. At a certain point, one's fingers simply know what to do when dialing. Put a personal phonebook to manual use often enough and it becomes redundant.

Today's approach to fulfilling this wish is to dispense with the physical phonebook in favor of a smartphone contacts folder. Superficially, this is a perfectly good and arguably more efficient and effective substitute. Having one's contacts on the phone is certainly simpler and more convenient than carrying around a separate phonebook. With virtual assistants like Siri and Viv, one can even circumvent the need to search one's contacts list. One can simply ask the phone to place calls. With the WeChat app developed by Tencent in China, even entering phone numbers into a contacts list is no longer necessary. It's enough to simply hold out phones and let them shake virtual hands and exchange numbers. Undeniably convenient, this solution to the problem of remembering phone numbers nevertheless has certain liabilities. When the bodily process of dialing phone numbers is eliminated, so is the basis of our intimacy with that number.

Phone numbers that are dialed for us never become part of our standard body–mind repertoire. Like walking up and down stairs, using utensils to eat, or tying shoelaces, remembering phone numbers and addresses is a simple action that we at first had to struggle intently to do successfully. No matter how many times our smartphone or smartspeaker places calls to friends or family members, we will never come any closer to remembering their numbers. People who move to a new city and use internet navigation systems to get around discover that, months later, they still need to use those systems. They do not internalize the city's layout. The city never quite becomes part of who they are.

Should it matter if we no longer remember phone numbers or addresses or how to get around in the places where we live and work, relying instead on our smart tools to do so for us? As long as we make sure to create backups, what difference does it make whether we do the remembering or our devices do so? This response seems reasonable enough. If remembering—whether phone numbers, addresses, or how to get around—can be reduced, without loss, to simple acts of storage and retrieval, what difference does it make if these acts are carried out electronically rather than organically?

But given that consciousness is an irreducibly temporal mode of relationality, might supplanting human remembering with digitally mediated processes of storage and retrieval result in an attenuation of consciousness? There is now strong scientific evidence that mind and cognition are best understood as embodied, embedded, enactive, and extended.[1] Although remembering may seem to take place "in our heads," it is not purely cerebral; it is *situated* and *embodied*. The same, of course, can be said of attending and learning—effortful practices that are crucial to intelligence. It seems increasingly evident that our minds are not reducible either to brain function or to some immaterial psychic essence and that they emerge in the action-oriented coupling of brains, bodies, and environments (Clark 1997). It follows from this that anything that decouples our brains and bodies from our environments risks compromising the qualitative depth and scope of our conscious, intelligent presence.

To begin envisioning where wish-fulfilling technology might lead, consider the mundane fact that infants are happiest when their mothers (or other intimate care-givers) are interacting with them with positive affect. If a mother begins ignoring her baby, failing to respond to laughs and coos, the baby will soon begin expressing distress. If babies and their mothers are linked via a two-way television loop, however, an interesting difference appears. If a baby sees live footage of his or her smiling and attentive mother, all is good.

But as soon as the televised image segues from live to recorded footage of the mother, even if the affective and attentive quality of the image remains the same, the baby will become visibly upset. What is crucial, it seems, is *relational immediacy*. It is only in live interaction or *active coordination* that we are able to enter into the shared temporal process of participatory sense-making whereby new domains of social meaning are generated (De Jaegher and Di Paolo 2007).

These observations invite pointed questions about the course set for the Intelligence Revolution by the algorithmic coding and reinforcement of choice, convenience, competition, and control as cardinal technological values. What are the implications of efficiency- and profit-driven campaigns to progressively offshore physical and intellectual effort to machine intelligences? Is it only jobs that will be lost? Or is something much more fundamental at stake? Will the ongoing development of emotional computing and social robotics make it possible one day to offshore care? What will happen as bodily effort and relational immediacy are elided, for example, from the practices of friendship and family? What outcomes and opportunities will result from the effortless satisfaction of our wants and the granting of our wishes by systems of ambient, artificial agency that are continuously learning, with superhuman vigilance, how to solicit and hold our attention? If intelligence consists in adaptive conduct—realizing new relational dynamics through learning- and memory-shaped responses to changes in our environment—how will our own intelligence be compromised as functional aspects of brain-body-environment coupling are "outsourced" to virtual substitutes?

Contrasting Scenarios for Futures Free from Want

Living in China today, it is not hard to imagine a near future in which one shuttles seamlessly from apartment to work via elevators and subways with minimal delay and with any street-level transportation arranged automatically and on time by the latest and smartest ride-hailing app. Already, many people prefer using food delivery apps instead of going to grocery stores or restaurants since they save time and are hassle-free. Shopping platforms like Alibaba offer same day or at most overnight delivery on virtually any consumer good, from running shoes to air purifiers. And for most young people, music and video have never been anything but streamed services.

For many of us, it might have been hard to imagine this way of life becoming normal, much less normative. But as was witnessed during the emergency response measures taken in most countries in response to the COVID-19 pandemic in 2020, meeting all of our basic needs without leaving home is entirely feasible. In the aftermath of that global experiment in "smart" living, including online education and telework, it is not implausible that movement toward the techno-utopian future promised by Big Tech will accelerate—a future in which automated production and delivery systems drive down the costs of goods and services far enough that work becomes entirely optional. In fact, it is not hard to imagine a future in which virtual realities become varied and veritable enough that spending as many waking hours as possible in them might be a welcome alternative to the actual living conditions of the global majority kept economically relevant through a universal basic income. Within a generation or two, it's possible that everyone might be provided with a basic living space of a standard size and layout that could be virtually altered on demand to have any décor and amenities one might wish. Travel might be possible but rarely necessary since it would be possible to share ideas, enjoy art together, discuss news and entertainment, and lead digitally stimulating lives, all in the safety of one's own home.

This is the "want-free" future imagined by the novelist E. M. Forster in his short novella "The Machine Stops." The story centers on a middle-aged mother and her young adult son. Like nearly everyone else, the mother lives in an underground apartment and spends her time earnestly hashing and rehashing the thoughts and doings of people she has never met in person. Her son, though, is a rebel. He has traveled on the surface and found direct sensory experience utterly captivating. He convinces his mother to meet face-to-face, but she is disturbed by his physical presence and frightened by his impassioned disenchantment with the beautiful lives provided to everyone by the Machine—the intelligent global infrastructure that people have come not just to respect but to revere as a beneficent, almost godlike presence. Learning of his exploits outdoors, the Machine threatens him with "homelessness" if he continues along his "unmechanical" path. But, as it happens, the Machine is in need of maintenance that no one any longer remembers how to perform. The story ends with the apocalyptic crash of the Machine.

Forster wrote this novella in 1909, a century before internet-based social media and the normality of endlessly repackaged web content and viral tweets. Two decades later, in the late 1920s, Nikola Tesla, the famous Serbian-American inventor of alternating electrical current and wireless energy transmission, envisioned a contrasting "want-free" future in which everything and everyone in the entire world is wirelessly connected through devices smaller than the human hand to form a single, electronic brain, resulting in a truly planetary intelligence—the birth of a world alive to itself in which each person benefits from the knowledge of all people, all of the time, leading a life of boundless convenience and choice.

Today, when Americans spend an average of just under twelve hours a day using electronic media and less than an hour outdoors, it is hard to deny basic validity to Forster's extrapolation of a future humanity content with exclusively virtual relationships. Tesla's lucent vision of boundless connectivity and choice, however, better captures today's dominant structures of feeling and Big Tech's promise of increasingly customized experiences and life options: freedom *from* prescribed identities and freedom *to* form wholly elective communities with whomever we wish. Rather than the grey totalitarianism and lost individuality envisioned by George Orwell in *1984*, we seem to be on course for a technicolor future of hyper-individuality. It may, however, turn out to be one in which lives organized around the values of convenience and choice ironically turn out to be lives in which choice becomes compulsory and in which every gain in convenience facilitates further attenuations of commitment.

The Future from Which Our Present Might Be Remembered

Forster and Tesla were each quite successful in extrapolating from their present to envision key characteristics of our present. Their successes were not based on seeing into the future through temporal telescopes but on being able to sail imaginatively on the winds of change at their own historical moments into futures from which their own presents might be remembered. That is our task as well. From what future(s) might we remember the world taking shape now in accord with the core values of efficiency, security, convenience, and choice, driven by competition among commercial, state, and military interests seeking to fully harness technological powers of control and sustain attention-fueled economic growth? What seems most imaginable as future norms for a sense of self, for learning, for sustaining health and maintaining security, and for friendship and family?

The Quantified Self

One definitive trend entrained with the economic and political imperatives for circulations of ever greater volumes, velocities, and varieties of data is an embrace of what has been called the quantified self. A brilliant and poignantly satirical vision of quantified selfhood in a society organized around the values of efficiency, security, convenience, and choice is presented in Gary Shteyngart's novel *Super Sad True Love Story*. Set in a near-future New York City and an America on perpetual war alert in which the government has been largely privatized and in which most people are employed in either Credit or Media, the novel explores the search for real intimacy in a world where power-brokering elites are intent on exploring the possibility of a post-human departure from mortality.

With a satirist's devilishly keen eye, Shteyngart sketches an "ordinary" way of life in which nearly all social interactions are conducted with the help of a personal "äppärätï"—a handheld device that functionally blends current smartphones features, the data-gathering and data-sharing practices associated with the "quantified self" movement, and a fully commercial algorithmic variant on China's experimental social credit ranking system. In Shteyngart's future America, everyone's äppärätï is constantly uploading time-logged data regarding practically everything they do—what they eat and drink, the quality of the air they have been breathing, the information they have been accessing—and correlating it with biometric maps of mood swings and arousal levels to provide everyone with up-to-the-minute calculations of their credit ratings, with alerts about nearby options for improving those ratings and with a running account of how their physical and mental performance is measuring up to age- and occupation-matched standards that are being set by their digital doppelgangers.

This kind of intelligence gathering and personalized feedback for the purpose of self-enhancement are well within current technical horizons. The self-tracking data being gathered by people in the "quantified self" movement today is being used for "body-hacking" to improve physical performance and to identify "creativity-boosting" correlations among activity, location, sleep patterns, and nutritional and sensory intake. In Shteyngart's future, these data streams are also used to establish compatibility baselines and hookup possibilities in a plausible extrapolation from what is already being done with mobile, location-based social search applications like Tinder and Tinder Social. But even more illuminatingly (and for most people, more depressingly), Shteyngart's fictional äppärätï also provide continuous updates on how desirable one is in the eyes of others. Simply point your äppärätï at another person and you are granted real-time data about how he or she is responding to you across a range of interest areas from sex to business opportunity, including data about how you rank in comparison with everyone else nearby. Of course, the äppärätï also enable you to instantly monitor how altering your behavior is affecting your desirability ranking and for whom. This is a future of obsessive, crowd-sourced, and algorithmically directed self-improvement.

Quantifying the self might seem innocent enough—a "cool" way of taking our data exhaust and using it empirically to facilitate our individual pursuits of health, wealth, and happiness. Although current concerns about data privacy might complicate movement toward such an openly data-saturated, expository society, China's experiment suggests that these concerns might be quite effectively evaporated if the benefits of buy-in are sufficiently radiant. Insurance companies in the United States are experimenting with outfitting policyholder vehicles with smart sensor systems that analyze driving habits and set premiums based on individual driver data, using machine learning algorithms to establish the exact price points needed to alter driving behaviors and maximize corporate profit. Avoiding automobile accidents is, of course, very much in every driver's interests, and schemes like this might well be regarded as "win-win." Yet, what if similar data-driven behavioral modification techniques were used to discipline other forms of risk-taking? What if the risks involved were intellectual or emotional?

For a vast number of digital natives, it is already the case that perceptions of self-worth are profoundly implicated in capacities for capturing and holding the digitally mediated attentions of others. If buy-in to a "social assistance" system like that envisioned by Shteyngart seems likely to facilitate being bathed in the attention of appreciative (and not critical) others, data privacy losses might easily be seen as no greater deterrent than the privacy losses entailed by marriage. The almost total loss of privacy involved in moving in and living intimately with someone is simply one of the conditions for deepening connection and commitment. One willingly ceases to be wholly individual to realize a mutual suffusion of one's self and another's. But for those whose self-identities are already digitally distributed and who may, in fact, see their digitally disseminated selves as primary, the kind of privacy associated with body-bound autonomy can easily be experienced as privation. Quantifying the self can amount, perhaps positively, to the computational expansion of the self.

It is easy to downplay the revolutionary character of movement toward a future of digitally mediated selfhood. Granted the presumption that we exist fundamentally as individuals, it is easy to equate the digitally distributed and the digitally disseminated self. As presumed individuals, the process of putting ourselves "out there" digitally is akin to that of sending out representatives of ourselves that can be engaged by others. But if we understand ourselves to be irreducibly relational, digital connectivity involves our *extension* not our representation. The internet is, in practice, a digital domain of relational opportunity—a domain of personal transformation. To see this, however, is to see as well that those who have the power to determine who connects with whom, how often, and for what reasons are in the position to shape who we are and what we mean to and for one another.

The realms of relational possibility being realized through intelligent technology are, as noted earlier, wholly intentional realms. The progressive quantification of self will, among other things, make it possible for those managing digital connectivity to suffuse their intentions among our own. This might take the form of invisibly imposing ideological constraints—a kind of soft coercion. But much more likely it will take the form of capitalizing on personal desires to optimize "win-win" outcomes and opportunities for all. Yet, as we will emphasize later in discussing ethics and social good, what is good for each and every one of us as individuals may not be good for all of us relationally.

Guaranteed Learning

In Shteyngart's future America, almost no one studies anything because "all" knowledge is instantly and effortlessly available on demand in attractively produced formats like YouTube videos and TED talks created in studios outfitted with every scientific aid available for assessing their effectiveness. No one even bothers to research what products to buy since marketing and retailing systems in both virtual and brick-and-mortar stores are continually updated by your äppäräti to be able to offer the most highly customized, credit-raising consumer options possible. You don't even need to carry purchases home. Quite literally, all you have to do is choose. Everything else is handled automatically.

This is not an inconceivable future. The quantified self-movement is still on the social periphery. Facebook, YouTube, Twitter, and Instagram ratings are not yet linked to one's financial, biometric, or medical data. The social credit systems being tested in China are not yet being used to their full, operant-conditioning potential. But at many American gas stations today, face recognition software and profile-generating algorithms are linked to cameras on the pump so that the advertising appearing on the monitor can be geared to your own demographic, based on your age, ethnicity, hair style, clothing, and vehicle type. Given current trends in smartphone and smartspeaker use, the mass migration from real to virtual social environments, and the explosive expansion of the internet of things, Shteyngart's future of knowledge apps and the obsolescence of formal education may well be in our shared destiny.

The prospects for transforming teaching and learning based on lessons learned from the digital remakes of marketing and retailing through intelligent technology have not gone unnoticed. Corporate experiments applying AI, adaptive gaming, virtual reality, and data-driven personalization to education have been ongoing for a decade and are ripe for scaling. AI and adaptive learning programs are being used, for example, to automate grading, to assist in producing quantifiable student learning outcomes, to personalize instruction in terms of content and pacing, to provide teachers with data about the parts of their lessons that students are not comprehending effectively, and to provide basic tutoring services. Gaming environments are being adapted for educational use, including virtual tours and role-playing activities at historical sites. And massive open online courses (MOOCs) are readily available that feature lectures by renowned scholars from some of the world's most elite universities. The incentives are quite clear. Annual education spending accounts for roughly 4.5 percent of global GDP or $4 trillion.[2]

Even so, prior to the COVID-19 pandemic, it was inconceivable that every student in nearly every country on Earth, from kindergarteners to doctoral candidates, would overnight be taking all of their classes online. That global experiment did not end up permanently shuttering schools worldwide. Its results have been highly uneven. But it has vastly accelerated the education technology (EdTech) movement and solidified its heading in the direction of increasingly individualized, digitally mediated learning.

Looking ahead, in anticipation of increasingly smart societies, education researchers are predicting a dramatic shift away from the centuries-old practices of school-based education centered on the standardized delivery of fixed sequences of learning tasks using one-to-many lectures, individual note- and test-taking, and direct peer competition. The envisioned future is one of intelligent educational environments in which students are guided through online knowledge content—for example, recorded and live video lectures, texts, immersive games, and virtual experiential learning environments—by automated teachers and tutors able to design custom learning opportunities that maximize acquisition and retention by using biometric and other feedback to assess and enhance student progress. The anticipated norm is to have individualized education available anytime, anywhere, and for anyone.

The learning environments imagined in science fiction books like Ernest Cline's 2011 novel, *Ready Player One*, are perhaps as little as a decade in the making. But the fact that major gaming companies—like the Chinese company NetDragon—are funding EdTech subsidiaries practically guarantees their eventual rollout. More far-fetched scenarios in which knowledge can be downloaded to the brain or in which the digital universe of knowledge is directly accessible via a brain–computer interface are generally considered scientifically possible but only minimally plausible. What seems undeniable is that in the same way that the Intelligence Revolution has made it possible to outsource memory and has transformed remembering into a commercially available service, the tendency will be for learning also to be made as sure and effortless as possible. Given the fact that technology giants like Cisco, Amazon, Google, and Tencent are investing heavily in educational initiatives, the corporatization of education is virtually assured.[3] Rather than a difficult but eventually rewarding practice, learning will be transformed into a service that will be available digitally and on demand.

Subscription Health

If we extrapolate further along these lines from our present moment, prospects become even more disquieting. Healthcare is another major domain for the expansion of smart digital services. The dramatic pretense of Shteyngart's future is that most people happily participate in the always helpfully assisted competition to raise their individual social and fiscal credit ratings. In fact, however, it is already clear that the structural inducements to engage in constant self-assessment on today's social media platforms are conducive to increased and intensified feelings of loneliness, anxiety, and depression (Davey 2016). Ironically, however, this might serve to reinforce and expand incentives to digitalize both physical and mental healthcare.

Smart devices can now be implanted that dispense medication as needed, based on doctor prescriptions and patient status updates. Smart implants are also being tested in the treatment of posttraumatic stress, depression, and eating disorders. There are compelling advantages in scaling up and broadening the use of this kind of technology. Using implants rather than pills to meet the medication needs of the elderly solves the problem of forgetting or confusing doses. In treating mental health disorders, implants make it impossible for patients to refuse or purposely skip doses—a major problem in pharmaceutically treating this class of health problems. Moreover, with smart implants, it is possible for doses and drug combinations to be adjusted as needed, based on constantly updated data, without the patient having to see a physician.

The dangers of prescription abuse and of self-medicating to relieve pain and to manage mood are well-known, and the use of smart implants offers an alternative. In principle, this could be seen as akin to pharmaceutically treating conditions like hypertension or obesity that are caused by a combination of lifestyle, diet, and stress factors. Although it amounts to treating symptoms rather than causes, pharmaceutical treatment is reasonably justified as preferable to allowing these conditions to persist untreated. With many illnesses and systemic ailments, symptom alleviation is not only humane, it is often a condition of possibility for effective treatment and healing.

As pharmaceutical precision increases, it does not seem particularly unlikely that mood and emotion regulation will one day be seen as either medical rights or as matters of technologically facilitated health choice, rather than as a result of long-term and effortful personal practice. Given the trend in Silicon Valley and other high-tech hubs to use micro-dosing of psychedelics to enhance creativity and improve performance without the major sensory and linguistic disruptions that often accompany use of these chemical compounds, it would not be surprising if machine learning systems are soon set to the task of identifying molecular compounds that will safely help users attain optimal functioning in a range of contexts. Building on recent successes the neuroscience of affect and cognition and in the use of noninvasive brain–computer interfaces to restore physical functionality to patients with movement disabilities, it is conceivable that adaptive brain-tuning will become a health and well-being option.

Yet, the most dramatic, near-term healthcare impacts of AI and the quantification of self are likely to be systemic. Brynjolfsson and McAfee (2017) are almost certainly right not only to expect a steady transfer of medical labor from human to AI experts across the care spectrum from diagnosis to treatment prescription, surgical intervention, prevention oversight, and research but also a breakdown of clear boundaries among them. As constant data generation becomes the norm, so will expectations of constant medical advice and intervention.

There is no reason, for example, why one's health data cannot be utilized to individualize the provision of preventative medicine. People today are using smartphone apps to remind them to relax during the day and take "mindfulness" breaks. But it would be quite easy to have one's eating habits regulated, for instance, by having a virtual personal assistant serve as one's dietary conscience, suggesting alternatives while grocery shopping. Less gently, one could simply be prohibited from buying groceries or ordering restaurant-prepared dishes that would increase one's health risks beyond some acceptable limit. This could be determined by individual choice, by healthcare plan providers, or by government mandate. Refusal to comply, depending on the societal context, could easily range from simply living with greater health risks, to being denied services, to paying higher insurance premiums or being held liable for treatment costs associated with health problems deemed avoidable.

Structurally, what seems likely given current change trajectories is that much as music and videos have become automated streaming services responsive to one's expressed tastes, day-to-day health provision will become a largely automated service in which scheduled doctor's visits will be a thing of the past since one's health status will be continually monitored, analyzed, and addressed. This gathering of medical intelligence and the algorithmic analysis of the resulting data might be welcomed by many, especially if linked to options for virtual doctor's visits and automated prescription delivery. Subscribing to a streaming medical service might be preferable to low-cost health provider plans under which nonemergency physician consultations often need to be booked weeks—and in some cases months—in advance. It's conceivable, as well, that using algorithmic explorations of big data and increasingly detailed quantifications of self will yield superior healthcare results, especially in the case of rare and so-called systemic diseases that are notoriously difficult to diagnose and treat. Healthcare technology might, in effect, become part of one's overall environment—a kind of secondary immune system, identifying and responding to threats before one is even aware of being infected, ill, or at risk.

As intriguing as this possibility might be, it is important to note that the digitalization of healthcare services is, at present, likely to scale up the prevalent conception of health as individual freedom from disease and the maintenance of normal bodily and mental functions and of healthcare intervention as a process of identifying the proximal causes of functional or structural irregularities and pain and then eliminating these pharmaceutically, biomechanically, or surgically. Yet, medical science and public health studies strongly support recognizing that health is an expression of allostatic or adaptive readiness at the nexus of interdependencies among physiological, psychological, social, economic, cultural, political, and environmental dynamics. In other words, health is a multidimensional expression of intelligence in action. Healing does not simply consist in "a return to previous somatic norms but also meaningfully heightened awareness and increased relational capacity" (Hershock 2006: 48ff). Indeed, a Buddhist conception of health—consistent with the pursuit of *kuśala* outcomes—is that healing should result in our being stronger, more flexible, more resilient, more attentively astute, and more actively attuned to qualities of interdependence than we were before injury or illness.

It is not at all clear that smart health services could—or will ever—be oriented toward enhancing relational health. Consider, for example, the use of mindfulness apps to address stress and enhance subjective well-being. By letting users know that various vital signs are indicative of the onset of an episode of intensified stress, and guiding them through a short meditation exercise, apps like this can successfully lower blood pressure and restore baseline respiration. They will not necessarily help people to become better at identifying and responsively addressing the relational sources of stress. Indeed, by outsourcing attentiveness to subtle bodily registers of relational malaise to machine intelligence, apps like these might ironically compromise the adaptive capacities of their human users.

At a social justice level, there is much to be concerned about if personal data is used by insurance companies and healthcare providers to discriminate among low-risk and high-risk populations, to offer premiums based on risk-aversion, or to penalize those who choose to ignore the "smart" advice they receive. But more fundamentally, there are real concerns that smart healthcare will consist in realizing the kind of health outcomes that can be delivered with intelligent machine assistance, without delivering as well the kinds of opportunities that will allow human beneficiaries to become more personally resilient and intelligently responsive. Transforming healing into a smart subscription service might well guarantee quantifiably good healthcare results. Whether they could contribute to the realization of superlative health outcomes and opportunities is another matter.

Smart Security

The full spectrum quantification of the self also opens troubling prospects for differently addressing problems with the body politic. The 2002 Stephen Spielberg film, *Minority Report* (based on a 1960s short story by Philip K. Dick), envisions an utterly secure and risk-free future in which crimes are algorithmically predicted and preemptively annulled before they take place. Nothing like that now exists or likely ever will. But police departments around the world are aggressively embracing data mining and analysis by machine learning algorithms and AI to engage in "predictive policing."

Making use of both historical data and so-called domain awareness surveillance systems, predictive policing currently has the relatively modest goals of identifying crime "hotspots," tracking the locations of known criminals, keeping records of people who associate with known criminals, and increasing police presence in areas identified as being at high risk of crimes based on a range of factors, including current weather conditions and breaking news. At their most successful, these methods have enabled police departments to predict, for example, where the next in a string of robberies will take place. More in keeping with the dystopian vision in *Minority Report*, predictive policing has also led to police officers paying visits to the homes of people who have not yet committed any crimes but whose patterns of association and movement suggest that they are at high risk of doing so. Delivering personal warnings about the kinds of punishment people like this would receive for crimes that algorithmic analysis has identified them liable to committing is seen by some police departments as "smart" preventative policing.[4]

Predictive policing has been criticized for biasing police attention toward (often poverty ridden) places and (often minority) populations that are already over-policed, with little strong evidence that it reduces crime. But virtually all metropolitan police departments in the United States are adding it to their arsenal, and there is no indication of this trend reversing, even in light of strong Fourth Amendment privacy rights challenges and pushback against tools like facial recognition systems. In countries like China, where no such legal protections exist, predictive policing and domain awareness are being employed with much greater intensity, merging data from mobile phones and credit cards with police records to identify and neutralize potential threats to public peace and state security. China's "preventative" mass reeducation (and effective incarceration) of Muslims in Xinjiang province over the last several years has been, among other things, a proof of concept for AI-enabled social control.

While predictive policing need not be paired with active surveillance, the logic of doing so is practically irresistible. The data dream of police and the security industry is total domain awareness—achieving tireless and ubiquitous attentive presence. While there is a long history of police and security use of closed-circuit television (CCTV) systems, it is only recently that advances in data storage hardware and visual analysis software have made it possible for CCTV systems to yield real-time gains in successful police intervention and security enforcement. Previously, CCTV recordings had to be viewed by humans. Although cities like London and Beijing ten years ago each had close to half a million CCTV cameras in operation, making systematic use of the millions of hours of footage these cameras produced every day was practically impossible. That is no longer true. Today, face recognition software is performing at roughly human levels but at speeds that are many orders of magnitude faster. CCTV footage from city-scale systems can now be mined and acted upon in near real-time.[5]

Cameras operating at street level, however, cannot see around corners, and even a million cameras in a major metropolitan area cannot yield continuous and total domain awareness. With these limitations in mind, the US military developed aerial camera systems for use during the Iraq war that afforded complementary, top-down views of combat areas. These systems are now being marketed commercially by private companies like Persistent Surveillance Systems that offer surveillance services which (according to its corporate website) are like Google Earth combined with digital stream recording capability. These systems use airplanes and drones to generate searchable, synoptic, aerial footage that covers dozens of square miles of surface area. The massive amounts of resulting high-resolution, round-the-clock footage can be stored for periods of up to several weeks and make it possible, for example, to follow vehicles and individuals involved in a known crime both forward and backward in time, seeing where they went afterward but also where they had been prior to it. The investigative possibilities opened by this are nothing short of astonishing.[6]

In countries where no laws exist to prevent it, this kind of comprehensive visual surveillance data can be correlated with smartphone global positioning satellite data, as well as with credit and debit card uses, enabling the continuous tracking of individuals of interest even where visual contact is lost. But as the COVID-19 pandemic has made evident, even in countries with strong privacy protection laws, extenuating circumstances can easily lead to lowering legal barriers to what would otherwise be considered invasive contact tracing. In the history of security-enhancing surveillance, it has seldom been the case that what is technically possible has not become tactically available.

With the growth of the internet of things, surveillance possibilities will continue proliferating. Web-connected smart cameras the size of coins can be attached to buildings, trees, traffic signs, bus stops, subway entries, and public trashcans. The uses of virtual personal assistants through one's smartphone to provide information and—at least for early adopters of Viv and similar services—to carry out tasks of varying degrees of complexity can be intercepted by both state and nonstate actors. In fact, any sensor-equipped smart device can be used for surveillance, whether by "angels" of the state or by its enemies. Every device that facilitates surveillance—whether for commercial or state purposes—also opens new security risks and vulnerabilities.

This does not apply only to securing business, homes, or national borders. It applies to every security domain from energy security to food security. As security technologist Bruce Schneier (2018) has pointed out in alarming detail, every internet-connected device from toasters to medical implants, autonomous vehicles, and power plants is hackable. The internet of things is not just a facilitator of conversational commerce, it is a ubiquitously pervasive field of security risks—an almost infinite "attack surface." Given the current status of corporate-state marriages in regard to intelligence gathering, every expansion of attack horizons establishes imperatives for further intensifying surveillance-for-security in an attempt to stay one step ahead of both state and nonstate hackers. One of the characteristics of technologies as complex relational systems is for their successes to carry them across their own thresholds of utility to bring about the conditions of their own necessity (Hershock 2006: 91).

Virtual Friendships and Intimacies

There is no longer any doubt that when it comes to abstract, logical reasoning and discovering useful connections across large data sets, humans are no competition for algorithmic intelligences. Even in the case of "outside the box" thinking of the kind involved in designing an asymmetrical racecar chassis or learning how to play undefeatable games of *go*, machine intelligence is already capable of what, a mere five years ago, would have been regarded as impossible-to-achieve superiority. If anything remotely like the current pace of development in machine learning and AI is sustained for even another decade, there are likely to be few, if any, purely intellectual pursuits at which machines will not comfortably equal or surpass humans.

The one type of intelligence at which it has seemed reasonable to think machines will never be capable of excelling is emotional intelligence. An AI might be able to recite all of Shakespeare's sonnets and provide a detailed a summary of all extant commentaries on each of them, but it will never truly understand the question Shakespeare poses, for example, at the beginning of Sonnet 16:

> But wherefore do not you a mightier way
> Make war upon this bloody tyrant, Time?
> And fortify your self in your decay
> With means more blessed than my barren rhyme?

Only a mortal, with a body worn down by the relentless chase of Sun and Moon, with skin and muscle slack in losing battle with gravity, could ever truly *understand* and not merely *reference* Shakespeare's scorning lament.

Or so the argument goes. In actuality, things now are not so clearly cut, and in the future they are likely to be even less so. The tremendous advances made in face and voice recognition software have vastly expanded the horizons of affective computing. By combining AI, deep learning, biometrics, psychology, and neuroscience, it has become possible to design and build machines that can recognize, interpret, correctly respond to, and simulate human emotions. The most proximate goal is to make human–computer interaction more natural and multi-modal: improving both the human user's experience and the system's performance.

A sense of the potential of this work can be gleaned by considering that it is now possible for a machine to predict—based on less than a minute of visual facial expression and body language data—whether two people will bond affectively over the course of their conversation. That is an impressive capacity for emotionally sensitive observational acuity, even if the machine itself doesn't have a sense of what it *feels* like to enjoy an affective bond. Machines are also now sufficiently capable of reading emotions to provide autistic children with real-time interpretations of what the people around them are feeling. And machines provided with appropriate biometric and behavioral data are now able to monitor stress levels, assess the effectiveness of advertising, and even predict the onset of depression prior to the appearance of objective symptoms.[7]

The more distant goal of affective and social computing is to be able to create socially intelligent artificial "agents" that can sense the affect, preferences, and personalities of their interlocutors; make socially acceptable decisions; and adjust their interactive style depending both on the specific knowledge domain framing the interaction (distinguishing, for example, the difference in social tone in doctor–patient and fitness coach–client interactions) and on the socio-emotional profiles of their human counterparts. Artificial "agents" have already been crafted with enough of these basic capabilities to serve effectively as health coaches and intake counselors and to portray different ethnicities and simulate different personalities in doing so.[8]

What we are witnessing today is a shift from selecting the gender, accent, and communication style of our virtual personal assistants and navigation systems from a menu of standard options to having these artificial agents figure out on their own "who" we want them to be for us and how. The next generation of artificial agents will not be generic presences in our lives. They will actively adapt to us, taking into account both our overt responses to them and underlying biometric cues to better understand and implement our intentions, doing so in ways that will maximize our affective resonance with them. These will be "agents" that will "understand" how we think and what we want. But more importantly, they will be "agents" that will care *for* us in ways that will convince us they also care *about* us.[9]

Simulating emotions is not the same as feeling them. But the fact that machines in the foreseeable future will only be able to represent and not truly understand and feel emotions may matter much less than one might suppose. The artificial strategy game player AlphaStar does not think in the way we think. But those who play AlphaStar in a game of StarCraft II have no doubt that they are matched against a "player" whose skill and creativity in the game equals or exceeds their own. Social interaction is not a finite game like chess or the real-time strategy games at which machines have taught themselves to excel. But social interactions are in general both rule constrained and goal oriented, and the deep learning methods that enable AlphaGo Zero to play unbeatable *go* or AlphaStar to play champion-level StarCraft II will eventually enable machines to interact socially in ways that will convince us that we are interacting with *someone*—a "social being" recognizably like us.

How far in the future this will be is hard to estimate. Social robots, both humanoid and nonhumanoid, are being developed that incorporate lessons learned from research in embodied cognition and that employ deep learning methodologies to engage, much as infants and toddlers do, in imitative learning. Just as we did, these robots are learning how to behave appropriately from the bottom up. They experiment, make mistakes, try again, and seek the reward of what amounts to social approval. In a sense, these robots are undergoing socialization in much the same way that we did growing up. The difference is that when one robot has learned to properly handle a given social interaction, that knowledge can be directly transferred to other similar robots, as well as to ambient machine intelligences. Scaling up social knowledge for computational intelligences is in principle no more difficult than scaling up mathematical knowledge. While only one robot or virtual assistant out of five might successfully learn the first step in a sequence of social moves, all five would subsequently be able to begin from step one in learning step two. Once one of these five succeeds, the other four (or an additional ten or twenty or two hundred) could join it going from step two to step three. The learning curve, in other words, can be exponential.

Over time, the capabilities developed through social robotics will be combined with those in affective computing. The results may never rival those envisioned in the *Bladerunner* films—androids that can only be distinguished from humans by trained experts. But even if social robots remain obviously artificial—like the care robots Paro (a robotic seal) and Pepper (a humanoid but clearly robotic "child") that are being tested now in Japanese nursing and assisted living facilities—this will not matter. They will be ever more convincingly human in their capacity to read and represent human emotions, and the evidence is that our desire for affective resonance will compensate for their behavioral and expressive shortcomings.

Through her long-term studies of human interactions with "relational artifacts," Sherry Turkle (2011) has made evident the surprising extent to which we are willing to attribute life and agency to artifacts that may possess only minimal expressive capabilities. Even behaviorally primitive toys like My Real Baby and Furbie that "require" attention and that "express" pleasure and pain through the rudimentary facial and body movements are remarkably effective in soliciting care. People interacting with these toys know that they are not really alive. But they nevertheless regard them as "alive enough" to warrant affectionate regard. There is no reason to expect that as the affective repertoire of machines expands, we will not be quite willing to "befriend" virtual agents "who"—through texts and conversation—are able to provide us with a sufficient sense of personal connection and emotional resonance to serve as surrogates in the collaborative formation of personal identity and sense of self. Over time, and informed by broad spectrum data derived from the quantification of self, it is almost certain that such virtual friends will earn sufficient trust that we will grant them the responsibilities of serving as both intimate confidants and advisors.

Among the most telling findings of Turkle's work with teenaged Americans is not just how willing they would be to accept advice from a virtual therapist but how strongly they would prefer doing so. Their logic was simple. Unlike real psychotherapists who have biases of various kinds and a limited knowledge of research in the field, a virtual therapist was perceived as unbiased and as having access to practically unlimited data on which to base its advice (Turkle 2011). Joseph Weizenbaum was, it seems, entirely correct to worry, even in the mid-1970s, about the readiness with which people were willing to attribute both care and psychological acumen to Eliza, his (by today's standards) conversationally inept chatbot. Our yearning to *be understood* runs deep, and we are—consciously or unconsciously—quite willing to set aside reasonable doubts about the capabilities of our interlocutors in order to have that experience.

Extrapolating along existing lines of development in affective computing and social robotics, it is easy to see a slippery slope developing between accepting virtual friends and confidents as "real enough" to serve as friendly presences to embracing them as welcome complements to, if not substitutes for, actual friends. This is quite convincingly depicted in the 2013 Spike Jonze film, *Her*, in which a socially isolated, recently divorced man develops a romantic attachment for a new intelligent operating system that he downloads onto his smartphone. Whether this would constitute a slip into misplaced affection or simply the emergence of a new kind of friendship or romantic relationship is, for now, not the critical issue. Developing a philosophically cogent understanding of the relationship between "actual" and "virtual" reality is something that we will need to do in due course. But the more immediate personal question is simply how actual and virtual friendships are likely to differ *as friendships* if the technological production of virtual friends continues developing along already established lines of research and interest.

Accepting the likelihood that actual and virtual friends will come to be experienced as *equally real*, it seems that there nevertheless will be at least two important differences between them. Actual friends will be physically as well as intellectually and emotionally present. That is, they will have an interactive presence in which we can share, consciously and subconsciously, and for which there will be no lossless virtual translation. Even if we allow that the communicative abilities of virtual friends might one day become indistinguishable from those of our actual friends, none of our virtual friends will relate with us through touch or on the chemical or "gut" level, both of which are fundamental to how we engage one another as embodied beings. In short, the *relational bandwidth* that we share with virtual friends will not be as broad as the relational bandwidth we share with actual friends.

How much this will ultimately matter, and whether the bandwidth of virtual encounters could be technically expanded to rival or exceed those of actual encounters, are open to debate. If video chatting is ultimately not fully satisfying as a medium in which to conduct a friendship or explore romantic potentials, it is likely that the narrower relational bandwidth we will enjoy with virtual friends will correlate with some sense of a shortfall or unfulfilled remainder in relational quality and depth. The extent to which this could be offset, for example, by the use of haptic devices that could place us in virtual touch with others or through chemical vapor generators that could emit body-matched pheromones and scents is something that undoubtedly will be explored empirically, as will more basic questions about the experiential equivalence of actually and virtually shared presence.

The more important difference is that while actual friendships require a great deal of effort to sustain and deepen, we have every reason to anticipate that virtual friendships will not. Making an actual friend typically begins with exploring mutual interests, sharing relevant past experiences to establish grounds for narrative convergence and compatibility, and then feeling our ways forward together toward deeper sharing. At some point, however, tension arises—a need to accept some significant difference—as well as times when one has to put aside one's own interests to be of help. The best friendships combine the pleasures of easily achieved mutual understanding with the sometimes very hard work of being wholly present *for* someone else.

Being present for others is, of course, at the heart of healthy family intimacies. Parenting practices and the social forms taken by familial relations have varied greatly over human history and across cultures. But all are arguably rooted in the cultivation of readiness to help and of caring attentiveness to one another in the context of both joyously and tragically shared life fortunes. Parenting is hard work, which was traditionally undertaken in apprenticeship under one's own parents and grandparents—the local experts to whom one was always able to turn for assistance and advice. Like all social and cultural institutions, however, the family is subject to change.

Smart Families and Digital Socialization

The first three industrial revolutions brought about a gradual but fundamental shift away from the centrality of work carried out within the home to employment outside the home. Most historians and sociologists today do not believe that industrialization led to the breakdown of the extended family. The nuclear family is not a product of industrial modernity. But as greater numbers of people joined the formal workforce, and as both supply and demand for mass-produced clothing, food, and household goods increased, along with needs for formal education, a host of activities like weaving cloth, sewing clothes, preserving fruits and vegetables, housekeeping, furniture building and mending, house maintenance, the transmission of basic skills, and healing were all transformed from commonplace domestic practices into specialized professionalized services.

For the most part, the mechanization, specialization, and professionalization of labor have been a good thing. We might harbor romantic attachments to the idea of growing our own food or making all of our own clothes, but most of us are better off letting the marketplace satisfy these needs, and the efficiency of mass production is hard to argue against in a world of nearly eight billion people. While complaints about education and healthcare are commonplace and often justified, there is little doubt that formal mass education and professional medicine have vastly improved our lives and that specially trained teachers, nurses, and doctors have knowledge and skills that most parents and grandparents do not. But just as we have reasons to be concerned about the ramifications of transforming learning and healing into smart services, there are reasons to be concerned about smart services expanding to include the most fundamental of domestic practices: parenting and socialization.

It has become common over the last several generations for radios, televisions, and video game consoles to be used by parents not only to entertain their children but also to introduce them to the world beyond the home in ways that are educational. These, however, are generic devices with generic programming. Intelligent technology holds out prospects of fine-tuning educational and entertainment media and providing highly personalized learning and entertainment services that are customized to meet each and every child's distinctive needs and tastes. Indeed, it holds out prospects for bringing global knowledge data and information resources to bear across the full range of parental decision-making and problem-solving domains. Smart family assistants, available through any sensor-equipped and web-linked device, have the potential to evolve into what amount to all-purpose *parental prostheses*—digital extensions of parents designed to serve as attentive and adaptive interfaces with their children. A smart family assistant could, for example, learn to wake children up and shepherd them through hygiene rituals and clothes selection in ways they enjoy. It could establish child-specific dietary regimes and meal plans, oversee refrigerator and pantry contents, and monitor vitamin and calorie intakes. It could help diagnose basic health problems, monitor sleep patterns, and recommend appropriate interventions based on each child's biometric data and a global database of best practices. It could take care of school and extracurricular transportation needs in conjunction with on-demand autonomous vehicles; recommend child-appropriate stories, books, music and films; and serve as an interactive and never-distracted virtual babysitter and teen counselor.

Most of us today might regard these possibilities with horror. But a hundred years ago, almost anywhere in the world, the idea of turning over the care of our aged parents to professionals in a nursing home would have been deemed equally unthinkable. The social valorizations of hyperindividuality, elective community, and unfettered choice are in sharp tension with the relational demands of achieving and sustaining domestic intimacy.[10] It is hard work making a family work. If some of that time and effort could be successfully outsourced, is that necessarily a bad thing? Young people who have grown up in smart environments interacting with and through relational artifacts are already more comfortable representing emotions than expressing them, are more concerned with behaviors rather than motives, and are increasingly prone to manifesting a deep "ethics gap" between belief and action (James 2014). Will they view the use of digital prostheses to carry out basic parenting duties as being any more problematic than using a smartphone to make, maintain, and break friendships and romantic connections?

We are on the verge of passing from the smartphone-wielding generation of "digital natives" to a smart service generation of "synthetically socialized" consumer/citizens who from earliest childhood will have lived in the company of virtual assistants and friends, whose synthetically intelligent teachers will have drawn almost magically on both global databases and intimate sociobiophysical knowledge of their students' learning responses, and whose spaces for play and work will have always been densely populated with ambient intelligences devoted to ensuring that they have experiences of the kinds and intensities that they want, peppered judiciously, perhaps, with opportunities for parameter-specified "personal growth."

Reliance on abstract systems rather than face-to-face relations may be a core characteristic of modern, highly urbanized societies, from which there may be little opportunity of pulling back (Giddens 1984). Virtual sociality may simply be a "logical extension of a culture that prioritizes 'mind' as the primary characteristic of being human" (Miller 2016). But if ethics begins in face-to-face encounters with people whose presences cannot be simply subsumed in our own (Levinas 1961), if virtual connection comes to be seen as an acceptable surrogate for full bandwidth relationality, and if the decoupling of embodied and social presence further intensifies, there will be costs in moral and emotional intelligence.

Considerable neuroscientific and psychological evidence now exists that asynchronous, digitally mediated connectivity compromises social learning, empathy, and emotional fluency by failing to activate the mirror neuron system (Dickerson, Gerhardstein, and Moser 2017). The mirror neuron system facilitates registering others' actions, emotions, and sensations in the same part of the brain at work when we experience these actions, emotions, and sensations ourselves. It is this perceptual-motor coupling that enables coordinated and dynamic perception-production cycles, not just in individuals but among them (Rizzolatti 2005; Hasson et al. 2012). In short, the predominance of digital learning and sociality is likely to produce significant social learning deficits.

Social learning is crucial to successful socialization, which is in turn crucial to social participation that effectively coordinates individual talents and capabilities. Any deficit in social learning, especially among younger children, will have potentially lifelong ramifications. Not only would digitally mediated socialization allow commercial values to be infused into the character development process, it would have the potential to compromise intentional attunement and the development of skill in realizing bodily and affective synchrony with others (Vivanti and Rogers 2014). As such, it has the potential to place at risk biosocial developments that are foundational for ethical deliberation.

Smart Services and the Sociopersonal Risk of Forfeiting Intelligent Human Practices

The future reach of the attention economy and of the colonization of consciousness is not yet determined. The Buddhist teachings of impermanence and karma insist on the liberating value of not thinking otherwise. Changes of values, intentions, and actions in the coming years have the potential to redirect the course of history. If sufficiently concerted, they will do so significantly. What we can be certain of at present is that the Intelligence Revolution is currently impelling the human-technology-world relationship in the direction of granting functional centrality to intelligent technology by fostering the contraction of attention- and effort-rich intelligent human practices and the expansion of smart services. These range from utterly banal digital photo manipulation apps to digital-first business services—offered by intelligent technology leaders like Amazon, Microsoft, Google, and Baidu—for everything from investments to customer service.[11] Across the entire spectrum, digital systems are being authorized to carry out actions and achieve desirable outcomes that would otherwise have been carried out by humans, perhaps less effectively and almost certainly less efficiently.

The appeal of smart services is undeniable. If time equals money, saving time by using digital apps and cloud services is saving money. That value is readily quantifiable. Some time and attention are spent when using digital apps and services as smart tools to get things done, but the returns on investment are clear. And if the apps and services don't deliver what we seek, we can simply stop using them. But if we zoom out to consider more fully what is involved, things are not so evidently cost- and risk-free. Even setting aside the data-generation side of the equation—the work done by the consumers of digital goods and services as producers of digital training data—attention spent selecting among smart service recommendations differs qualitatively from attention directed toward actively working toward desired outcomes oneself. In effect, what digital services make possible is the translation of human practices into digitally executed actions. There are risks involved.

As noted earlier, the first and most basic of these human practices—remembering—has already been transmuted into a digital memory storage and retrieval service. This is not surprising. Even Babbage's mechanical difference engines had to be able to store and retrieve data to serve as mathematical savants. But to be incapable of remembering is to be incapable of adaptive conduct. The most significant difference between conventional and autonomous vehicles is not really technical (the addition of more electronic sensors, for example), it is temporal. Memory renders time sensible. But through making causality-sensitive anticipation possible, it also makes time malleable. Skill in remembering is thus crucial to both responsibility and responsiveness, and it is significant that while the English word "mindfulness" is used to refer one of the most basic Buddhist meditation techniques, the original term (Pali: *sati*; Skt: *smṛti*) actually means "memory," and the practice of mindfulness is at root remembering things—literally, putting things back together in interdependent relationship. Without memory, it is impossible to purposefully change the way things are changing.[12]

While we are not at risk of entirely abandoning the practice of remembering, our willingness to use digital memory services is already causing most of us to become increasingly "out of practice" with it. If the Intelligence Revolution continues on its current trajectory, the same will become true of other core human practices like learning, healing, securing ourselves against harm, and caring for others. This is no small matter. Remembering machines are also capable of purposefully changing the way things are changing, and they are purpose-built.

The deskilling of humanity through the forfeiture of intelligent practices like remembering, learning, and healing may take some time. In spite of the impressive feats of machines with "deep learning" capabilities, machines still do not learn as well as humans except within tight and well-defined parameters. But even without achieving parity with humans, the incorporation of AI and machine learning within our institutions for education, healthcare, and security will radically alter the character of our personal presence in these domains. It is not only cybersecurity that is being outsourced to AIs and robots, it is also the physical security of homes and businesses. Robotic "agents" tirelessly process data to identify actual and potential threats. They are lightning fast in their responses, and so the same logic that legitimizes an AI and robotics "arms race" in the military domain will at some point also legitimize the transformation of smart security services from a possibility to a norm. There is nothing in this logic to keep this from becoming true in the domains of educating and healing as well.

Supplementing and even supplanting these human practices with or by machine services may not seem like something to resist. To learn more efficiently, to be healthier, and to be more secure with the help of machines is something most of us are likely to embrace, just as we generally have embraced the modern institutions of schools, hospitals, and police forces. Bringing machine intelligence to bear on these domains makes apparently unassailable good sense. What child would not gladly forfeit educational struggle for effortlessly and playfully acquired knowledge and skills with the help of a smart learning service? Over time, however, translating these effortful human practices into machine services will change what we *mean* by learning, healing, and securing, and this will come with risks. Technologies change what is regarded as possible and preferable, shaping conduct conceptually as well as behaviorally.[13]

Current developments in affective computing and social robotics leave little room for doubt as to whether smart services are likely to spread as well into other domains, placing at risk such distinctively human social practices as making friends and caring for family members. These will not be existential threats of the kind that might be generated by the creation of an artificial superintelligence. They will not directly threaten humanity's continued physical existence. But the practice-to-service transition in these domains could very well place our humanity at risk, threatening our commitments to and capacities for social intelligence and truly humane becoming.

It is arguable that what is most distinctive about humanity is our seemingly unmatched collective propensity for opening new domains of agency and experience—new spaces for the exercise of intelligence. These include the aesthetic domains opened by dramatic, musical, and artistic practices; the epistemic domains opened by scientific practices; and the intimate relational domains opened, for example, by befriending and parenting as practices of belonging. Other animals may be capable of minimal presence in these domains. But no other animal enjoys anything remotely like the effortful reach—both extensive and intensive—that we humans have realized in the domains of culture, science, and intimate sociality.

Opening each of these domains has involved a distinctive expansion of human consciousness: an expansion of the relational dynamics in which we participate constitutively. The relational potential in each domain is infinite. Their realized dimensions and characteristics, however, are limited and shaped by the consistency and quality of the effortful practices through which we establish our active presence in them. In the absence of effortful practice, we can enjoy some experience in these domains. But to be present appreciatively in the strong, dual sense of both valuing and adding value to these domains, we must have worked hard enough to develop appropriate kinds of adaptive, improvisational agency. We must have developed capacities for expressing and exercising intelligence within them.

Musically arranged sounds, for example, can be heard effortlessly by any living being equipped with the functional equivalent of ears. But to go from hearing musically arranged sounds to listening to music requires attentive effort, including the effort of sustaining a continuously remembering presence. To explore music through singing or playing an instrument requires, in addition to effortful listening, other kinds of effortful practice. Ultimately, there is no creative *musical agency* without effortful and intelligent *musical practice*.

AIs are now able to compose original and emotionally rich music and create unique works of visual art.[14] They do not, by any means, rival human artists in their ability to do so. But their abilities will improve. The deep learning systems in Netflix's algorithmic recommendation engine that are learning how best to capture and hold our attention can be turned to producing rather than predicting satisfying video content. A future in which AIs supplant the artistic practices of authors, film directors, and musicians is perhaps distant but not implausible (Tegmark 2017).

It might be objected that this does not need to be seen as a zero sum game. The transition from push plows to livestock-pulled plows to mechanical tractors and finally to smart tractors and harvesters has been marked at each step by increases in productivity and savings in human time and labor. The technological demise of hand cultivation not only enhanced food security, it enhanced quality of life, freeing people to use their time and energy in things other than back-breaking toil. Why should the technological demise of hand-played music or hand-painted art be any different? If artificially intelligent smart services colonize the entertainment industry, wouldn't that free up human time and energy for other, perhaps even more profound, pursuits?

These objections cannot be summarily dismissed. But they are premised on having some kind of common "currency" in terms of which one can quantify what is both lost and gained, for instance, in the translation of traditional agri-*culture* into industrial agri-*business*. By using a stable, tangibly measured value like food productivity to assess this shift, it's possible to discount or entirely overlook such variable and intangible values as intimate partnership with the land or the diversity of climate- and geography-shaped cuisines. Agriculture and agribusiness are conducive to realizing profoundly different environments for and qualities of interaction among plants, animals, and humans. Similarly, if the merits of smart service tools are assessed on the basis of a single metric like user-satisfaction, what relational possibilities do we stand to lose and how might this loss affect qualities of human consciousness and intelligence?

The Predicament Ahead

We are now in a position to state the global predicament into which we are willingly being ushered by the technological transformations of the Intelligence Revolution. Commercial, state, and military interests have set into motion desire-inducing and wish-fulfilling adaptive machines, fueled by our own attention energy, that are crafting environments designed to elicit, interpret, and respond to our intentions with inhuman speed and precision in order to extend and intensify the powers afforded by competitive advantage in the colonization of consciousness.

If as persons, communities, corporations, and nations we continue acting as rationally self-interested individuals, changing the way things are changing in accord with headings fixed by triangulating among the values of choice, convenience, and control, we will become willing parts of a system for "automatically" getting precisely and only we want. Unfortunately, to get better at getting what we want will embroil us in becoming better at wanting and thus in being never quite able to want what we get. That karmic loop will tighten. Either we change who we are becoming or we will become ever more deeply subject to wish-fulfilling objective structures from which we will be unable and unwilling to exit. Our subjectivities and the objectivity of those structures will be related like the impossible-to-separate interior and exterior of a Klein bottle. We will become that which contains us, enjoying practically unlimited individual freedoms of experiential choice in ever more compellingly controlled environments.

Building this logically beautiful, karmically closed system is the most ethically troubling and existentially risky endeavor on which humanity has ever embarked. Today's intelligence industries are not devoted to making paperclips. Neither are they malevolently directed at eliminating humanity. What they exhibit are extraordinary potentials for bringing about conditions in which choice becomes a compulsive end-in-itself and in which humane commitments to humane sociality, diversity, and equity will be almost surgically eroded. Fortunately, it is human attention energy and the information carried along with it that are keeping the computational factories of the intelligence industries humming. Without our attention energy, the data grid would lay idle and the servants, savants, seers, and sirens of the Intelligence Revolution would be dark and silent. That gives us collectively enormous responsibility. Who do we need to become to accept and act on that responsibility most aptly and ethically?

5

Anticipating an Ethics of Intelligence

Identifying the attention economy and the colonization of consciousness as engines of the Intelligence Revolution makes evident the need for an ethical counter revolution. As the art of human course correction, ethics involves us in the intelligent practice of evaluating values, intentions, and actions to realize better lives as persons in community with others. Determining which relational headings are better or best is seldom a simple matter. Ethics often requires prioritizing values, discovering how to bring competing goods into productive alignment—skill in the art of predicament resolution.

Just as there are many navigational systems, there are many ethical systems. If we are going to responsibly and responsively change the course of the Intelligence Revolution, which ethical system or systems are best suited to understanding and evaluating its dynamics? More specifically, which ethical orientations are most consistent with developing and sustaining the kinds of consciousness and attention that are at the heart of ethical deliberation and predicament resolution in environments shaped by systems of domination, not through coercion but through craving satisfaction and through optimally presented options for choice?

It is no longer necessary to argue on behalf of the importance of bringing ethics to bear on developments in artificial intelligence (AI), machine learning, and big data. Academic, corporate, and governmental centers and initiatives dedicated to doing so are being established and growing at seemingly exponential rates. But, for the most part, the approach being taken is to make use of ready-to-hand ethical systems to address what are taken to be essentially issues of agency. Broadly speaking, this means raising and addressing ethical concerns about risks associated with *accidents of design* and about *misuse by design*. That is, ethical concerns tend to be directed toward technical shortcomings like algorithmic bias resulting from skewed training data in otherwise benign uses of AI or toward bad actors with such malign intentions as using autonomous vehicles to deliver improvised explosive devices.

What remains much less common is ethical concern about the *structural* risks of these technologies—risks of the kind that occur, for example, when the benignly intended affordances of smart services alter the environment of human decision-making and social intelligence.¹ What kind of ethics is best suited to understanding and addressing the unprecedented structural risks of the Intelligence Revolution?

Ethical Possibilities

It is plausible that reflections on what should and should not be permissible and on the meanings of and means to a good life are as old as human communities. For much of human history, however, these reflections seem to have been immediately focused. They were reflections on qualities of lives in *this* community and the actions of *these* people. The transition from community-defined deliberations on morality ("our" headings/ideals) to explicitly general humanity-enhancing ethical deliberations ("any" reasonable person's headings/ideals) seems to have occurred first in the cultural ecotones or zones of cultural interfusion that began emerging with long distance trade and urbanization roughly twenty-five hundred years ago on the Mediterranean coast, on the Indo-Gangetic plain, and along the Yellow River in what is now China. Yet, as is clear in comparing the resulting ethical systems, their aspirations for generality did not keep them from being regional. Classical Greek, Indian, and Chinese ethics are as distinctly flavored as their music and cuisines.

It is useful to keep this in mind in considering which ethical system or systems might be best suited to countering the inhumane potentials of the Intelligence Revolution. In spite of their sometimes vehemently proclaimed universalism, until very recently, all ethical systems have been products of reflections on regionally relevant issues and have rested on conceptual sediments of still earlier deliberations taking place within localized moral communities. They have not, in other words, been truly global ethics developed in response to manifestly global issues through the deliberations of an all-inclusive global community. If this is so, and if the issues that are being raised by intelligent technology are indeed unprecedented and global in both origin and character, then this arguably mandates the development of an entirely new and thoroughly global ethics (Górniak-Kocikowska 1996).

Claims about the uniqueness of the ethical challenges posed by the new intelligence technologies can be traced at least to the mid-1980s, when James Moor argued that digital computers and information and communications networks were not only making previously impossible actions possible, they were actualizing these new possibilities before any substantial ethical deliberation had taken place as to whether these new kinds and domains of action *ought* to be developed. This is a crucially important observation. In

Moor's memorable phrasing, by putting new modes of action into practice before they could be considered ethically, these new technologies were both creating "policy vacuums" and revealing "conceptual muddles" (about agency, for example) that until then had been concealed by the narrow horizons within which concept construction had been carried out (Moor 1985).

Today, the computational factories of the Intelligence Revolution are crafting environments in which human agency is being dramatically shaped and reshaped by our interactions with artificial and synthetic "agents" whose conduct could never have been the subject of any prior ethical reflection. Neither these computational agents nor the environments within which they are exercising their autonomy existed even ten years ago. It is thus both surprising and significant that—more than a quarter century after claims began to be made about the need for a new *kind* of ethics—professional philosophers are still debating whether AI is simply making visible ethical issues that have "been there all along" or if it is raising truly new issues (Boddington 2017). It is useful to pause a moment and ask why that debate persists.

Part of the answer is that, for professional philosophers, it's tempting to assume that the longevity of existing, mainstream ethical systems is evidence of their perennial validity. Moreover, while their historical origins might have been regional, the fact that they are now taught around the world can be interpreted as evidence of their global maturity. Whatever ethical issues the Intelligence Revolution might raise, resolving them should be possible by making use of one or another of these time-tested ethical systems. Even if these issues should turn out to be truly unprecedented, the corollary assumption is that it should be possible to rehabilitate existing ethical systems to address them.

In fact, all of the mainstream, centuries-old ethical systems—virtue or character-focused ethics, deontological or duty- and principles-based ethics, and utilitarian or more broadly consequentialist ethics—are now being actively used to engage issues emerging with intelligent technology. This has been especially evident in the field of robot ethics and in deliberations regarding standards for the design and engineering of evolutionary algorithms and autonomous vehicles.[2] But as the ethical guidelines developed by the Institute of Electrical and Electronics Engineers (IEEE) and the Association for Computing Machinery (ACM) make evident,[3] these mainstream ethical systems remain what amount to default alternatives.

The list of existing, originally "regional" ethical systems that might be used to address the global ethical challenges of the Intelligence Revolution—either as is or in some rehabilitated form—can, of course, be expanded to include, for example, pragmatist, continental, or feminist ethics, as well as various

Asian ethics traditions.[4] But the eventual failure of a rehabilitation strategy is, I think, already evident in the precedents for them existing and persisting as distinct lineages of ethical practice. Each of these ethical systems has ethical blind spots. As their critics have pointed out, "tough cases" afflict each system: cases in which applying that ethical system leads to counterintuitive, patently unjust, or morally unacceptable outcomes.

The classic example is how the deontological principle of only telling the truth becomes morally repugnant when you are asked by Nazi investigators if you are harboring any Jewish people in your home. If you are doing so, according to a strict interpretation of the duty to tell only the truth, you should admit as much. Should you tell the truth, however, the neighbors you have been safeguarding will be taken away and will likely be subjected to mortal hardship, while your own family may also face potentially life-threatening reprisals. Comparably "tough" cases also plague the systems of virtue and utilitarian ethics. Indeed, it is these "anomalous" issues and circumstances that in part explain the crafting and continued existence of competing ethical systems. Each ethical system has readily established strengths when it comes to addressing the "tough" issues and circumstances that afflict other systems of ethics.

In sum, the plurality of existing ethical lineages can be seen as grounds for their *individual disqualification* as global ethical systems. Even if one or another of them could somehow be sufficiently freed from its historical regionalism to serve as a universal or "global" ethics, it would still have its blind spots. Applying any such ethics globally would serve only to expand the real-world compass of those blind spots. Given this, the obvious question is whether it would be better to build a new global ethics from the ground up—an ethics purpose built to engage the distinctive concerns about values and humane becoming that are being raised the Intelligence Revolution.

Toward an Ethics of Information

The first attempt to develop a built-for-purpose global ethical system to address the unique and troubling issues raised by the Intelligence Revolution or its integral technologies was made by Norbert Wiener (1894–1964), the "father" of cybernetics. In *The Human Use of Human Beings: Cybernetics and Society*—written in 1950 at the dawn of contemporary efforts to conceive and create AI, and in the midst of escalating competition between the Soviet Union and the United States as aspiring global hegemons—Wiener issued an impassioned call for demilitarizing the emerging technologies of electronic computing and information processing and for directing their development toward human, rather than inhuman, uses of human beings.[5] In Wiener's terms, justly developed computing and information technology should help ensure that all humans have the freedom to realize their full potential, that all are equal in the sense that judgments about the justice of a situation would not change if our positions in it changed, and that all people will act benevolently, at every scale from the most personal to the global, emphasizing considerations of humanity as a whole.

Wiener was not a trained philosopher, and his reflections present a morally motivated plea for humane technology rather than a general and formally articulated ethical system. What proved to be most influential was the ontological underpinning of his plea: a cybernetic conception of life as entropy-reducing information generation and processing. In Wiener's view, living organisms and computers are information systems operating, respectively, on biological and electronic substrates. In each case, the complexity of their structures is an index of what can be expected of them in terms of engagements with their environments: an index, essentially, of their varying capacities for improbable and yet nonrandom conduct (Wiener 1950: 57).

This broad, entropy-reducing conception of a system's potential has been used more recently by Terrell Bynum (2006; 2007) to develop a global ethics that blends Wiener's cybernetic insights with Aristotelian notions of flourishing. Bynum's "flourishing ethics" seeks to resolve three anomalies or shortcomings inherent to traditional ethical systems: their tendency to negate all other ethical theories, their affliction with troubling cases they are unsuited to addressing, and their difficulty in extending ethical consideration to nonhuman agents. By conceiving of flourishing as excellence in doing what one is equipped to do, "flourishing ethics" becomes sufficiently capacious to encompass concern for the well-being of all sentient organisms. But, in addition, it extends ethical consideration to the well-being of entities that may possess sensory capacities and means of acting in and on the world without being alive. It is a global ethics, then, in the sense that it eschews preoccupations with solely human concerns to consider "the broader, and more reasonable, goal of the flourishing of life, ecosystems and just civilizations," as well as that of "well-behaved cybernetic machines that participate in the very fabric of those civilizations" (Bynum 2006: 170).

In Bynum's view, "flourishing ethics" does not replace but rather opens possibilities for deepening understanding of traditional ethical systems. This is consistent with his convictions about the value of pluralism, but it is a point that he does not elaborate on sufficiently to see precisely how traditional ethical systems relate to his putatively global alternative. Indeed, in light of the first anomaly attributed to traditional ethical systems, there is an implicit irony involved in constructing a global ethics that is itself singular—a free-standing, individual ethical system instead of, for example, an ethical collective. This is especially true if this global ethics is colored by Aristotle's explicitly teleological understanding of flourishing, which implies that change is naturally and most appropriately directed toward an already existing, consummate end or goal. Using Wiener's more open-ended notion of order as entropy-reducing informational conduct to offset Aristotle's teleology unfortunately has the liability of simply eliding—rather than resolving—the toughest ethical issues: those that require prioritizing equally compelling but presently conflicting values and interests. In the language introduced earlier, it has the liability of using a single metric for quantifying differences in values, intentions, and actions, thus effectively translating ethical predicaments into technical problems.[6]

Consider, for example, the order or flourishing that would result from the algorithmic customization of online experience to maximize attention capture for the purpose of compressing the gap between desire generation and satisfaction, thus accelerating consumption and overall economic activity while individualizing the pursuit of happiness in ways conducive to reducing dissatisfaction with the prevailing political system. How would this compare to the order or flourishing that might result if the human experience was not subject to the actions of algorithmic agencies and if, instead, half of the attention energy currently devoted to consuming electronic media was used aesthetically to enhance our own local life conditions? If the total entropy reduction or information generation in these two cases was somehow quantifiably "the same," would their difference fail to be "informative" enough to distinguish "ethically" between their very different developmental scenarios?

Patient-Oriented Ethics of Information

One way of addressing this concern is to introduce greater complexity into the concept of flourishing and to focus on ethical relationships rather than agents and actions. This is the approach taken by Luciano Floridi in his philosophy and ethics of information. Like Wiener and Bynum, Floridi takes information and entropy reduction as central concepts on which to build a global ethics that is capable of offering guidance in evaluating and shaping the future course of computing, information, and communications technologies. In his early work, Floridi laid a radical ontological foundation for his ethics by claiming that "to be is to be an informational entity" (Floridi 2008a: 199). Assuming that existence is preferable to nonexistence, it follows both that if existence is information, then "all entities, qua informational objects, have an intrinsic moral value" (Floridi 2007: 10) and that the degradation or elimination of information is intrinsically bad. The only "evil" in the world is Non-Being or what Floridi alternatively refers to as "metaphysical entropy" (ibid. 200).

Floridi later modified his ontology to stress the generation—rather than the simple existence—of information, stating that "to be is to be interactable" (Floridi 2013: 10). This marks an important relational qualification of both agency and goodness that is consistent with his arguments for focusing primarily on ethical patients, rather than ethical agents and actions. Emphasizing ethical patients—that is, those who bear or receive the consequences of action—effectively subordinates considering the moral status of individual actors and actions to considering experiential and situational outcomes. This shift of ethical focus is especially important in addressing the distinctive issues of agency that are raised, for instance, by algorithmic "actors" and ambient intelligences. Potentially, it also opens prospects for addressing the structural—rather than misuse and accident—risks of deploying machine intelligences.

Traditional ethical emphases on individual (human) agents and their actions afford very little critical traction when it comes to responding to the ethical challenges of distributive agency, the behavior of global actor networks, or the adaptive conduct engaged in by AI systems, all of which are part of our contemporary technological reality. Seeing persons as interactable informational entities addresses this shortcoming. In addition, it sheds usefully different light on the meaning, for example, of our presence on social media platforms and on the capitalization of data gleaned from that presence. If we are informational beings, our social media presences are not *representations* of ourselves; they are *extensions* of ourselves. This has profound implications for how issues of privacy, surveillance, and intellectual property are understood, legally as well as ethically.

Floridi's affirmation of the basic goodness of the cosmos, his endorsement of universal respect for any entity, his identification of being with interacting, and his redirection of ethical regard toward ethical patients are aimed, at least implicitly, as affirming ethical plurality. Rather than a "view from nowhere" that applies a single standard of evaluation to all things, Floridi presents his ethics of information as a "cross-cultural platform" upon which to engage in shared reflection—a "neutral language" for uncovering different conceptions of ethical issues without having to commit to any particular culturally laden position (Floridi 2006: 113). Whereas actor- and act-oriented ethics typically identify universal standards for what counts as ethical, specifying the type(s) of agents and actions that are morally valued, this is not the case in patient-oriented ethics since the uniqueness of each informational entity is presumed foundationally.

Questions might be raised, of course, about the ultimate cogency of a purportedly neutral language and platform. In addition, and more immediately worrying, is the fact that Floridi's embrace of plurality and his core principle of universal respect for the intrinsic goodness of all entities—including cultural and institutional entities—can be seen as an endorsement of ethical relativism. As informational entities, should torture chambers really be accorded the same respect as animal research laboratories or as hospital surgical theaters? Few, if any of us, would go along with such an "egalitarian" view of the moral value of these very different institutional frameworks for interaction among sentient beings. Floridi agrees and makes use of the concept of levels of abstraction to clarify his position.[7] In the case of these three spaces of action, there are a number of shared levels of abstraction at which we might compare their value or moral worth: architectural, historical, financial, legal, emotional, epistemological, and so on. Depending on the level of abstraction chosen, we might have quite different value rankings of the three. Which level of abstraction or set of levels we *should* employ would depend on the goal of our analysis.

Although there is considerable merit to the method of levels of abstraction, it is not clear that it is an adequate response to worries about falling into "anything goes" relativism. As James Moor has pointed out, "values saturate our decision-making" (Moor 1998: 18). This includes our decisions about which level of abstraction is most relevant given a specific goal as well as our decisions about which goal is most appropriate in a specific endeavor. Entropy reduction simply does not seem to be the kind of value that could guide these decision-making processes with appropriate critical acuity. Moor's response to charges that his own endorsement of ethical plurality amounts to an affirmation of ethical relativism is to call attention to the fallacy that according validity to many ethical alternatives amounts to affirming the equal acceptability of any and every one of them. His conviction is that the value-saturated nature of life, combined with the fact that all human beings and communities have certain basic needs and interests, allows us to identify a set of core values or goods found across all societies that can serve as a common baseline for evaluating ethical alternatives. We can effectively qualify, in a usefully universal manner, how an informational difference obtains among situations that are otherwise equivalent in their entropy negation. Among these core values and goods, Moor includes life, happiness, health, security, freedom, resources and opportunity.

The shortcoming of this solution for keeping the affirmation of ethical plurality from collapsing into an endorsement of ethical relativism is that there is little historical evidence for persistent and common core values. On the contrary, there is considerable scientific evidence that concepts are always both historically and culturally conditioned (Huebner et al. 2010; Park and Huang 2010; Machery 2004; Weinberg 2001). The conceptions of even such socially basic phenomena as family and friendship vary radically across cultures, and deep disparities are apparent in how values like freedom are conceived, for example, in Islamic, Buddhist, Daoist, and modern Western thought. This shortcoming is profoundly amplified in the context of the ongoing, technologically mediated transfer of attention energy from human practices to smart services—a transfer that, as noted earlier, is already changing what is meant by goods/values like education, health, security, friendship, and happiness.

It is one thing to *proclaim* the value of ethical and cultural plurality and another to think, speak, or write in ways that *demonstrate* ethical and cultural plurality. Indeed, such proclamations, coming as they do from one or another single perspective (even if jointly authored), are arguably incoherent in the sense that they necessarily fail to be what they endorse. They are, in other words, only able to pay (perhaps quite reverent) philosophical lip service to plurality. This liability is, however, not just a matter of method—unilateral authorial expression rather than conversational elaboration, for instance— but also one of goals. The desire for a single principle, standard, or universal matrix of common values and goods against which to measure the worth of different perspectives on values, intentions, and actions, and on the experiential (patient-received) outcomes and opportunities associated with them, is evidence of a deep-seated modern bias toward the ideal of unity. This is not only out of sync with contemporary realities; it is inconsistent with the recognition of and respect for differences in identity and history that are the necessary first steps toward valuing and not merely tolerating plurality.

A truly global ethics that refuses the absolutism of a single principle or set of core values and at the same time rejects "anything goes" relativism must, at the very least, shift focus from seeking common values to engaging in the improvised realization of *shared* values in the sense delineated by Jean Luc Nancy (2000). For Nancy, the "common" always carries implications of discipline or coercion, as it did in Nazi invocations of the "common" values of the German people in the decade leading up to the Second World War and Holocaust. A "shared" set of values is not one that (against all the evidence of science and history) is somehow the same in and for each of us. Instead, it is a set in which we each a have distinctive, contributory stake: a set of values that incorporates and enables the activation and further articulation of our differences. What we are in need of is not a *common* global ethics but a truly *shared* global ethics. Or, stated in more metaphorical terms, to address the challenges of the Intelligence Revolution, we do not need a new *species of ethics*; we need an *ethical ecosystem*.

An information-based global ethical ecology may, in actuality, be a fair characterization of what Wiener, Bynum, and Floridi have intended to advocate. But without specifying what characterizes an ethical ecosystem, it is not clear why forfeiting the search for ethical unity would be good for ethics. After all, we already have ethical variety. There are not only the three major lineages of traditional ethics—based on the primacy of virtues, duties, and consequences—but many subfamilies within each, as well as now well-recognized alternatives like pragmatism, care ethics, and environmental ethics. Would there be any ethical value in simply increasing their number, adding to the mix, for example, ethical systems reflecting the distinctive value constellations of indigenous or first peoples, of Asian or African cultures, or of different religious traditions? Adding more musicians and instruments to a musical ensemble does not necessarily make for better music. Similarly, there is no apparent necessary relationship between greater ethical variety and either better ethical reflection or more sure and certain translations of ethical theory into ethical practice.

Diversity and an Ethics of Intelligence

One way of specifying what is involved in realizing an ethical ecosystem is to extrapolate from the dynamics that characterize natural ecosystems. In ecosystems, what matters most is not the sheer number of species present but the kind and quality of relationships they share. To bring the nature of ecological relationships into better focus, let me stipulate a distinction between the variety and diversity. Variety consists in the bare presence of multiple things, beings, or processes. Diversity consists in a distinctive quality of relational dynamics.

More specifically stated, variety is a purely quantitative measure of the factual extent of *coexistence*. It is visible at a glance and can readily be mandated or imposed. In contrast, diversity is a qualitative index of the depth of *interdependence* present. It is a quality and direction of relational dynamics that emerges as differences are engaged as the basis of mutually reinforcing contributions to sustainably shared welfare. In short, rather than being a quantitative measure of *how much* things *differ from* each other, diversity is a qualitative index of *how well* things *differ for* one another. It cannot be seen at a glance and cannot be imposed. Diversity emerges only with the appreciation—the increasing value—of difference.[8]

To use a concrete example, in a zoo, the species from a savannah ecosystem merely coexist. They are only externally related and thus depend for their survival on outside inputs, including water, food, and medicine. Successful zoos exhibit species variety; the relationships among species are completely contingent. In a natural savannah ecosystem, species are interdependent and their survival depends on sustaining an equitable pattern of both direct and indirect mutual contribution. The relationships among them are internal or constitutive. Species diversity is thus an index of the quality of interdependence in an ecosystem. Or, in other words, diversity is an index of ecosystem health—the capacity of an ecosystem for resilience and adaptive responsiveness in the face of disruptive forces and conditions. More strongly stated, *diversity is an index of ecosystem intelligence*.

This understanding of diversity allows us to shed some useful light on the ways in which Wiener, Bynum, and Floridi have deployed the concept of entropy. Unlike the *fact of variety*, the *fact of diversity* implies the presence of responsive differentiation: the presence of mutually entailing processes of differentiation from and differentiation for others. That is, the factual presence of diversity is inseparable from the presence of diversity as a value—an improvisational modality of appreciation: the presence of *intelligent interdependence*. Entropy increases whenever the diversity present—for example, among cells in an organism, among organisms in an ecosystem, among cognitive approaches in a research team, or among cultures in an ecotone—reverts to mere variety. Entropy is intelligence drain.

Restated in terms consistent with this characterization of diversity, Floridi's assertion that "to be is to be interactable" becomes an affirmation that "to be is to differentiate interdependently." Or, going a step further: "to be intelligently interdependent is to differentiate responsively and responsibly." Differentiating responsively is what distinguishes adaptation from simple mutation. Differentiating responsibly is implied by the very notion of interdependence, strongly interpreted, since acting on others in a world of strong interdependence is simultaneously acting on oneself. Strong interdependence is an internal or constitutive relationship. Weak interdependence is an external or contingent relationship.[9]

Using this robust conception of diversity, it's possible to distinguish clearly between affirming global ethical plurality and realizing a global ethical ecology. Unlike ethical plurality or variety, ethical diversity implies the achievement of strong, constitutive interdependence based on appreciating ethical differences as resources for the progressive elaboration of shared values. Affirming global ethical plurality conserves ethical information. Yet, in and of itself, ethical plurality neither expresses nor fosters ethical intelligence. Realizing a global ethical ecology will entail initiating and sustaining responsive and responsible ethical differentiation that enhances philosophical adaptation and resilience in ways that are interculturally valued.

Some Buddhist Clarification

At this point, introducing the Buddhist concepts underlying this conception of diversity is helpful in distinguishing between an ethics of information and an ethics of intelligence as either candidate or collaborating approaches to addressing the predicaments posed by the Intelligence Revolution. Strong interdependence is what is implied by the interrelated Buddhist concepts of conditioned-arising (*pratītyasamutpāda*), impermanence (*anitya*), and emptiness (*śūnyatā*) or the absence of any fixed essence or identity. Elaborating on the relationship among these concepts, the Chinese Buddhist philosopher Fazang (法藏, 643–712) argued that given the absence of fixed essences and identities (emptiness), interdependence necessarily entails interpenetration or the dynamic (ever-changing) nonduality of all things.[10]

Contrary to familiar reductionist understandings of nonduality as either a "vertical" integration of the divine and the mundane or a "horizontal" integration of all things as part of a single substance, the nonduality invoked by Fazang is one in which all things are the same precisely insofar as they differ meaningfully from and for each other. That is, realizing the nonduality of all things consists in realizing that all things exist in an ecological matrix within which each particular (*shi* 事) at once causes and is caused by the totality: a world in which each thing *is* what it contributes functionally to the patterning articulation (*li* 理) of that totality. Simply stated, each thing ultimately *is* what it *means* to and for others.[11] Moreover, since to mean something to or for another is to make a significant difference to or for them, *to be* is *to be valued*. But this begs asking: valued as what and in what way, with what moral valence?

This Buddhist understanding of meaning-generating interdependence and interpenetration has apparent resonances both with Floridi's claim on behalf of the inherent value and goodness of everything and with his appeal to levels of abstraction. But a Buddhism-inspired ethics of intelligence differs in some significant ways from an ethics of information that is based on the ontological primacy of information/interaction and on the ethical primacy of ethical patients as informational entities, each of which possesses a unique *telos* or goal. Briefly exploring their differences will perhaps help clarify the value of ethical diversity.

In the terms just introduced, interaction entails weak interdependence. Actions are, in Floridi's terms, "messages" that are "inherently relational" in the sense of connecting otherwise independent senders/agents and receivers/patients. Whereas strong interdependence entails that our connections are constitutively meaningful, weak interdependence implies that behavior—that is, messaging—is only contingently meaningful.

Meaning occurs only in reception or interpretation. Thus, moral value cannot depend on an action's motives or consequences but only on how it affects some specific patient (Floridi 2013: 79). Any actions that allow an informational entity to flourish—that is, to fulfill its informational *telos* or destiny—are morally good; any actions that have the entropic effect of compromising an entity's informational integrity or degrading interactive possibilities are morally bad.

But if all things are inherently good, it becomes necessary to explain how informational agents and their messages/actions could have entropic or morally adverse impacts on informational patients. Recall that it is this problem that opens the ethics of information to charges of relativism. Floridi's appeal to levels of abstraction is a response both to these charges and to the related question of what the value or meaning of a message/action is contingent upon. The levels of abstraction method shifts the locus of moral value from the relationship between the agent and the patient to the patient alone and the locus of meaning more generally from the dynamics of interdependence to an independent process of determining the proper goal of interpretation in a given instance.

The method of shifting levels of abstraction roughly parallels the epistemic uses of the Buddhist concept of emptiness as a way of disarming presumptions of fixed meaning. In both cases, there is a denial of inherent value or meaning and an affirmation of their ultimately relational nature. But the metaphysical context is quite different. Meaning emerges *within* strong interdependence. Meaning is not, in the Buddhist case, dependent on propositions or judgments formulated *about* interdependence from positions supposedly independent of it. Meaning is relational inflection.

From this Buddhist perspective, seeing information/existence as ontologically basic and as opposed to entropy/nonexistence establishes a perspective from which interdependence can only appear weak. It involves projecting into what is present the characteristics of our assumed relationship to it as knowing subjects who are independent of objects known. Interaction is what remains once individual 'things-relating-to-each-other' have been abstracted from ontologically prior, internal, and constitutive relational dynamics. That is, interaction is predicated on individual things having been brought into existence as literally "standing apart" from one another. This act of abstraction is karmic in the sense of being undertaken intentionally, in keeping with some (set of) value(s). As such, it will result in some characteristic and responsibility laden pattern of outcomes and opportunities. The interacting entities found at any given level of abstraction are real. But, from a Buddhist perspective, reality is not something we discover; it is something we confer.

To put this somewhat differently, in a karmic cosmos, every experienced reality is the always negotiable result of different ways of inflecting relational dynamics. Every event has a moral valence traceable to our own decisions or, more accurately, our own histories of decision-making. From this Buddhist perspective, what ultimately exists is neither information nor entropy but openness, indeterminacy, ambiguity: emptiness (śūnyatā) as the limitless potential of relationality as such (tathātā). The dynamic interdependence and interpenetration of all things (admittedly a conventional, not ultimate, description) is not intrinsically good or bad. But this is something we realize only through our sentient capacity for inflecting relational dynamics in ways that are conventionally experienced as better or worse. What we do changes what things mean, but that in turn changes who we are since we *are* only what we *mean* to and for others.

As sentient beings, we are constantly disambiguating what is present. Seeing things *as* this or *as* that—seeing, for example, other animals as food or as pets—implicates us differently in irreducibly shared presences and relational futures. We create the conditions of our own intelligence or adaptive conduct. All too often, the result is duḥkha: environments conducive to experiencing conflict, trouble, and suffering. All is not good, all is not liberating. Goodness is not intrinsic to things or inherent to the cosmos as a whole.[12] It is not that the cosmos was good until we messed it up—the story of Adam and Eve in the Garden. Prior to intelligent differentiation (vijñāna, literally, "divisive knowing"), neither good nor bad existed.

This is something readily apparent in digital virtual realities—experiential environments that are built entirely from scratch. In them, nothing exists until it is purposely coded into existence. Responsibility is associated with everything present, even if some aspects were generated by randomization algorithms. If a so-called virtual reality is one that can be changed by those entering and acting in and on it, then it will transform in accord with the evolving character of agencies exercised therein.[13]

From a Buddhist perspective, however, at cosmic and evolutionary time scales, this is all that has been going on all along. Through intentional and value-expressing effort, the infinite potentiality of the suchness (tathātā) or dynamic emptiness (śūnyatā) of all things has been disambiguated into possibility-constraining and yet creativity-evoking realities. What we normally refer to as reality is the result of intentionally realizing certain actionable possibilities and ignoring or attempting to delimit others. Day-to-day life takes place in virtual or provisional (saṁvṛti) realities, worlds that we have enacted, and not in the "ultimate" (paramārtha) reality of unconditional suchness. They are realities provided we know and accept, for instance, the disambiguation of teachers, students, and staff members as aspects of relational dynamics in the wholly intentional environment of a school. What it means to be a good student or school is contingent on still further provisionally differentiated patterns of relationality. The Buddhist ethical question is how do our differentiations—our provisos about what counts as real and ideal—differ qualitatively?

Virtuosity as Ethical Orientation

The merits of Buddhism's cosmic vision can be debated. The important point for us here is that the meaning of goodness is not fixed, either for any sentient being—as defined, for instance, by its *telos* or purpose in existing—or in general. In early Buddhist traditions, the purpose of ethics was not understood to be guidance in leading a good life in the way we might suppose, based on typical understandings of "good" and "goodness." To use the early Buddhist term, ethics is the practice of leading *kuśala* or virtuosic lives—engaging in superlative thought, speech, and action and thus realizing virtuosic conduct or patterns of dynamic interdependence.

Virtuosity is, of course, relative. But it is not relative in the abstract sense implied by "anything goes" relativism. Musical virtuosity is relative to some existing standards of musical performance and consists, minimally, in exceeding those standards. Musical virtuosity reveals new musical domains. But to be deemed virtuosic, a performance—jointly realized by the performing musician(s) and an appreciative audience—must be experienced as extending performance horizons in ways that then establish new standards. Virtuosity is neither a private possession nor a public goal to be arrived at once and for all. Rather, *virtuosity consists in the effortful achievement of excellence within a community of practice, thereby setting new standards of excellence*.

In traditional Buddhist teaching contexts, the interrelated concepts of interdependence, impermanence, emptiness, and nonduality are introduced to foster *kuśala* engagement in personal and communal practices for reducing conflict, trouble, and suffering. The ultimate point of seeing dynamic relationality as ontologically more basic than things related is not theoretic; it is therapeutic. Realizing the nonduality or strong interdependence of agents, actions, and patients (or things acted upon) is not a metaphysical goal; it is support for the sustained practice of ethical intelligence. Ethical virtuosity is not something achieved *through* the practice of ethical intelligence but as achievement *of* that practice.

If intelligence is adaptive conduct, realizing *ethical virtuosity* as an *achievement of the practice of ethical intelligence* means engaging in conduct that exemplifies superlative capacities for and commitments to differing responsively and responsibly both from and for others. It is, in other words, practicing and fostering ethical diversity. Conceived along these Buddhist lines, an ethics of intelligence invites critical engagement with dimensions of the Intelligence Revolution that otherwise might not be deemed ethically salient.

The changes being brought about by the Intelligence Revolution may be good, for example, in terms of bringing about positive and measurable learning and healthcare outcomes. From the perspective of an ethics of information like that forwarded by Floridi, if these outcomes can be achieved in ways that respect our own integrity as informational entities and that allow both human and nonhuman agents to be held accountable and responsible, then they can be deemed ethically sound. Likewise, expanding cyberspace and crafting new and ever-differentiating virtual realities will bring into existence new informational domains within which to exercise ever-multiplying kinds of agency and in this sense can also be deemed ethically good.

To look slightly further ahead, if some parental practices and responsibilities—bedtime storytelling, helping with homework, arranging transportation to school and extracurricular activities, and ensuring healthy eating and exercise habits—can be turned over to virtual personal assistants, this also could be good. With access to all the resources available through the internet, a virtual personal assistant will have access to more varied stories and tutoring resources than any single set of parents ever could. Having a self-driving vehicle handling transportation will free many hours of parental time for higher level informational interactions than those involved in routine drives to soccer. With constant biometric monitoring providing feedback, a virtual personal assistant will almost certainly offer better daily health and exercise advice than all but the most highly trained human professionals, attuning itself to each individual's needs. In sum, as smart parenting services become the norm, parents and children may both end up measurably "happier" and "healthier" and will have more time for informational and experiential pursuits of their own choosing. Family members could all feel much freer in framing and satisfying their desires than in family life prior to smart services, especially with access to practically unlimited virtual and augmented reality options.

From an ethics of information standpoint, there is no clear technological limit as to how "good" lives can be if the Intelligence Revolution continues as it is. Worries may be appropriate about whether commercial, state, and military interests could use the powerful "digital pull" capacities of intelligence-gathering and wish-fulfillment networks to advance their own agendas through invisibly and irresistibly exercised "digital push." But these are worries about "bad actors" taking advantages of an otherwise good system, *not* about the moral valence of the system itself. In the end, if we are generally happier, healthier, better informed, and more secure, and if we enjoy greater freedoms of choice than ever, how could the system conferring this reality on us all be deemed anything other than good for each and every one of us, both as patients and agents?

A Buddhist ethics of intelligence invites asking whether such a system is able or likely to help us realize virtuosic relational dynamics and superlative social institutions. The economics of the Intelligence Revolution are premised on an intelligence-gathering infrastructure that enables AIs to identify, predict, and respond to individual human needs and desires. Increasing "personalization" is a crucial dimension of the seductiveness of smart services and the ultimate key to their commercial success. And given their tireless learning capabilities, if we assess these services ethically in terms of their benefits to us as individual agents and patients, we can expect to find them earning increasingly favorable evaluations.

Social good, however, is not simply an aggregate of individual good. To maintain that what is good for each and every one of us is good for all of us is to fall prey to the fallacy of composition. Using an automobile to get around is good for each and every one of us in terms of choice, convenience, and control—we go exactly where we want, when we want, with minimal physical exertion, protected from inclement weather, and so on. Yet, if everyone drives cars, we all end up stuck in traffic jams in cities with scarcely breathable air. The differences in life quality in "walking cities" and "driving cities" are evidence, among other things, of how strongly interdependent technological and social environments are—how things like frequency of significant chance encounters, depth of neighborliness, and the vitality of locally owned and locally operated businesses are environmentally conditioned.

If, as Buddhist ethics suggests, we see all things as strongly interdependent, ethical regard shifts necessarily from the interactions of individual agents and patients to relational dynamics. The primary focus in evaluating smart services ethically is thus not, for example, how individual family members fare when these services are brought into the home but rather how family structures and dynamics are affected. Whether smart services and the Intelligence Revolution are conducive to leading virtuosic lives is ultimately a question about how they affect our capacities for and commitments to relating freely and exemplifying ethical intelligence.

Outsourcing decisions about which restaurant to go to for dinner or what music to listen to are not, perhaps, matters of significant ethical gravity. But is the same true of decisions about what stories our children hear, watch, and read? If ethical intelligence is an achievement of the practice of differing responsively and responsibly from and for others—in order to not only alleviate personal conflict, trouble, and suffering but also to contribute to the realization of superlative relational dynamics—within what kind of familial environments are children most likely to engage in and sustain this practice? If we do not learn how to realize the kind of adaptive conduct (intelligence) needed to resolve predicaments arising in familial contexts, what likelihood is there of us doing so across cultural and national boundaries? If ethical virtuosity is an achievement of the practice of ethical intelligence and if smart services systematically supplement and eventually supplant intelligent human practices, will the first generation to grow up digitally socialized also be the last generation to demonstrate the effortful resolve needed to contribute to realizing a truly diversity-rich global ethical ecosystem?

Ethical Resolution

The ethical resolution that will be needed to address the predicament at the heart of the Intelligence Revolution on its current heading will not be realized by drawing on the resources of any one ethical system. It will require effortfully cultivating ethical intelligence and diversity. The intelligence-gathering infrastructure of the Attention Economy 2.0 is powering a complex, desire-inducing, and wish-fulfilling system that interprets and responds to our intentions with inhuman speed and precision, extending and intensifying the reach of the technological values of control, choice, and convenience, while at the same time bringing about social environments that offer few opportunities or incentives for practicing compassion, kindness, and attentive care. Resisting its seductions will require global as well as personal resolve.

Our relationship with technology is not merely interactional, a relationship of weak interdependence; it is a coconstitutive relationship of strong interdependence. Technologies do not just mediate or connect us with one another and the world. They implicate us in remaking both ourselves and our world as parts of a human-technology-world system (Rosenberger and Verbeek 2015). In designing and implementing new technological systems, we are at the same time designing and implementing new norms and processes for being (or becoming) human (Stiegler 1998; 2009). Long before any technical "singularity" occurs, through creating and using intelligence industries and the smart services they make available, we will have transformed ourselves and the meanings of both selfhood and sociality.

Although it is a staple of dystopian science fiction, the worry is not that we will spend more and more time in digital environments that constrain our interactive possibilities. In fact, it is only a matter of time before our interactive capabilities in virtual realities will exceed those that we have in actual reality. We will be able, for example, to perceive and act on events at scales from the microscopic to the planetary, making use of information gleaned from across the entire electromagnetic spectrum. With robotics, this reach will be extended into the actual world.

The worry is also not that we will be seduced into trading actual for virtual reality. If reality is not something discovered, but conferred, the actual/virtual distinction does not necessarily map onto the real/unreal distinction. Operating at a purely biological level, atoms and molecules are only virtually real. Without theoretical and technical assistance, we cannot perceive or interact directly with them. The physical structure children enter on the first day of formal education is a school *building*, but the cultural institution we refer to as a school exists at a level of abstraction to which children initially have no access. The *school* is a virtual reality that only becomes actually real over time. As social, cultural, and political agents, we are continually interacting in what began as virtual realities.

The ethical worry is that the Intelligence Revolution is currently on track to normalize our immersion in environments in which relational dynamics are recursively intent on expressing the normativity of control, convenience, and choice—realities in which intelligent human practices are at risk of becoming redundant and in which engaging in values-evaluating course correction seems destined to become a lost art. In considering how best to generate the critical resolve—the clarity and commitment—needed to reorient the Intelligence Revolution, Buddhist ethics would have us focus on three interrelated dimensions of this process: individuation, desire, and karma.

Individuation. The founding Buddhist insight was that alleviating and eliminating conflict, trouble, and suffering begins with investigating the interdependence of all things. This insight is now well-supported by biological and ecological science. But there is also growing critical engagement with the liabilities of living as individual, "bounded beings" (Gergen 2009). As the Confucian thinker Henry Rosemont Jr. has carefully and eloquently argued, the individual self is descriptively a fiction and prescriptive individualism—while it has contributed to recognizing and beginning to redress historical injustices in relation to gender, race, and ethnicity—is now proving to be a hindrance rather than a help in securing the conditions of truly humane community (Rosemont 2015).

Under the normal circumstances of daily life, it is fairly apparent that our autonomy is quite limited and that our well-being is intimately tied to that of others. We experience times of energetic and even ecstatically intimate harmony with others, but we also often find ourselves interacting with others who do not share our desires or who are impacted negatively by actions we had taken to be entirely unproblematic. A great many of the conflicts, troubles, and suffering experience, undesirable as they may be, are also valuable reminders of the liabilities of acting in ignorance of our interdependence. Productively engaged, they can be lessons in adaptive conduct: occasions for cultivating and expressing responsive virtuosity.

The technological systems involved in the Intelligence Revolution offer attractive pathways around these lessons. They have been designed explicitly to craft possibilities for us to behave *as* individuals, offering personalized portals into spaces in which we can choose to act as we want and experience what we wish as autonomous, rationally self-interested agents. Indeed, the iconic machines of the industrial and Intelligence Revolutions have been designed to scale up and accelerate the rate at which we individually exercise freedoms of choice—in consuming goods, in seeking pleasure, in using resources, and in connecting with others. The smartphone, the present state-of-the-art tool for attention capture and intelligence gathering, epitomizes the technology of individuation: a device through which our individual identities are continuously being mediated, made, and remade.

There are great benefits in being freed from both prescribed and ascribed identities. Yet, to the extent that this freedom is technologically realized, it is a freedom conditioned by the implicit values and constitutive force of the media through which we are making and remaking our identities. What seems to be the freedom to become whoever we want is, in fact, the freedom to become whatever is consistent with the values of the technological environment within which we are making and remaking ourselves. These values affect not only *who* it is possible for us to become but also *how* we do so and *why*. The freedoms of identity associated with digitally mediated sociality are derivatives of the contingency of connection and relationship therein, and this profoundly conditions relational quality.

As was noted earlier, while digital social media afford nearly unlimited options for membership in communities formed around exchanging life experiences, the reciprocity achieved comes with little or no responsibility. Moreover, the heightened connectivity options they afford do not correlate with deepening commitments of the kind required to put moral and social values into practice. Rather, what social media and habitual texting and messaging foster is the emergence of the "flattered self" as experiential sovereign, along with tendencies to identify the optional with the optimal (Zengotita 2005). None of this is conducive either to making consistent efforts to differ *for* and not just *from* others, or to developing skills for or commitments to shared predicament resolution. On the contrary, it is conducive to living in bias-confirming "echo chambers" (Del Vicario et al. 2016).

Desire. The growth of the intelligence industries depends on an evermore precisely and powerfully incentivized trade in desire. In sharp contrast with the bleak and choice-curtailing surveillance state envisioned by George Orwell in *1984*, smart capitalism and state surveillance make use of data gathered through commercial platforms that promote hedonistic indulgence and expository excess—a form of sociality celebrating the expression and gratification of personal desire. In this "expository society," algorithmic search and recommendation engines, smart streaming services, predictive analytics, and tailored advertising are being employed as tools to almost surgically disinhibit consumers and accelerate desire turnover.

Desires are not all bad. Although Buddhist ethics is sometimes associated with the eradication of desire, the Middle Way proposed by the Buddha was, among other things, a way of moving oblique to the spectrum of possibilities between ascetic denial of desire and hedonist indulgence in it. One of the most basic Buddhist meditative practices is to attend to how thoughts and desires arise and, after identifying their type and quality, to return to being simply and openly present. In early Buddhist traditions, desires were classified into three broad categories. The most troubling are desires based on self-magnifying cravings and attachments. These can be cravings for sense pleasures, for acquiring certain identities or kinds of status, and for getting rid of or avoiding certain things, people, or experiences. All of these are referred to as *tṛṣṇā* (Pali: *taṇhā*). But, in addition, there are *chanda* or desires to accomplish or pursue something with effort and resolve. This is the kind of desire that parents have when they strive to provide their children with opportunities to prosper in life and that children have when, in turn, they work hard to succeed in school. It is the kind of desire that one has in committing to learn how to speak a new language or how to play a musical instrument. The moral valence of these desires depends on the intentions and interests expressed in seeking to satisfy them.

The only desire that is deemed superlative or virtuosic is the desire for all sentient beings to attain enlightenment or liberation from conflict, trouble, and suffering—a desire for the responsive virtuosity needed to induce others to appreciate the enlightening potentials present within their own situations. This is a desire for relating freely that is rooted in the compassion of the bodhisattva vow and that can be fulfilled only through resolute practice: a desire for realizing dynamically intimate and liberating nonduality.

The intelligence industries are now directed toward constructing environments that will maximize attention capture and accelerate desire turnover. While participation in these environments does not prevent *chanda* or effortful desire, it is not particularly conducive to it and is arguably antithetical to actualizing bodhisattva desire. These are environments responsively attuned to possibilities for further inculcating *tṛṣṇā*. What will happen when even the minimal effort involved in wishing for things is no longer necessary because ambient AIs will have already anticipated our desires and responded accordingly? The process of accelerating desire turnover depends on being able both to gratify desires as quickly as possible and to ensure that satisfaction as fleeting as possible. What will the moral result be if we are never in a position to discover deficiencies in our intentions and desires because we only experience their fruits fleetingly before being enticed into new experiences by tireless sirens that have been coached by seers privy to everything we have ever said, written, done, and felt significantly enough for it to leave an informational trace?

Trial and error is not the only way to learn. But as a way of refining and reorienting our conduct in pursuit of less troubling or conflicted outcomes and opportunities, it is often the best—and sometimes, the only—way. An undesirable experience online can be curtailed instantly and replaced, which then becomes data for honing the skills of smart services. The flip side of getting better at providing desirable experiential options is getting better at preventing encounters with those that prove to be undesirable. This, however, amounts to algorithmic agents taking responsibility for learning what should and should not be desired—an offshoring of perhaps the most basic kind of moral and ethical effort.

Karma. Especially in global popular culture, karma is synonymous with destiny. But the Buddhist teaching of karma has nothing to do with our futures being somehow predetermined or fixed. In fact, karma is precisely what allows us to have confidence in being able to change where we are currently heading, based on our past history of making value-laden decisions and acting intentionally. As noted earlier, "karma" refers to a process of continually negotiated consonance among the patterns of values, intentions, and actions we actualize and the patterns of outcomes and opportunities we experience as both persons and communities. Because we can change our values, intentions, and actions on the basis of experienced outcomes and opportunities, we can change our life circumstances and the meaning—the relational ramifications—of our presence in them.

Smart services and search and recommendation tools, sharpened by deep learning algorithms, are crafting for each of us ever more personalized and enticing experiential "bubbles": virtual environments deeply attuned to providing us with everything we wish, and *only* what we wish. As such, they are becoming intermediaries for our karmic individuation. I get to experience what I like. You get to experience what you like. There is no need to give or be given feedback on whether our likes mesh. But there also is no opportunity to benefit directly from differences in what we like. As this system matures, we will be induced to spend increasing amounts of time and attention energy in what amounts to karmic isolation. This will not be like isolation in a physical prison. Even if the gratification of our desires is often only digital, we will not have the feeling of being in any way deprived. On the contrary, karmic isolation will be lusciously seductive: existing interactively in friction-free informational environments that offer no moral resistance whatsoever.

Once digital reality has sufficient sensory bandwidth, our experiential bubbles will be spaces akin to those envisioned in Hindu and Buddhist cosmologies as the abodes of the gods—heavenly spaces (*devaloka*) in which no desire will ever be denied. But, in Buddhist traditions, these heavens are understood to be unsuited to the practice of compassionate and enlightening effort on behalf of all sentient beings. The abodes of the gods are, at best, resting places prior to reentering the realms of karmically fraught conflict, trouble, and suffering within which enlightenment is possible. The implications for the meaning of Buddhist enlightenment or liberation are profound. But more to the point of our present discussion is the fact that the emancipatory subordination of heavenly realms to the human realm is also a subordination of the personal ideal of living as a god to that of living as sentient beings who cannot live without effort but who can vow to conduct themselves as bodhisattvas and embody the truth of relating freely. The compassionate, improvisational genius of the bodhisattva develops only in relational domains characterized by resistance and friction, in communities of practice in which there are standards of conduct and thus potentials for virtuosity. Adaptive conduct, our intelligence, languishes in the absence of challenge.

Clarity, Commitment, and the Value of Ethical Diversity

The great ethical traditions each bring a specific kind of clarity to understanding the origins and nature of the predicament of intelligent technology. Each has much to contribute to framing commitments for realizing the conditions of truly humane (and not merely human) flourishing. The same can be said of other more recent and broadly motivated ethical approaches like feminism and pragmatism, as well as of ethics purpose built to address issues raised uniquely by digital technologies.

But, in addition to singular, even exemplary practices of ethical intelligence, developing the clarity and commitment needed to resolve the complex predicaments of the Intelligence Revolution will ultimately require ethical diversity. Thus, the question with which we began, "Who do we need to present as in order to resolve global predicaments?" comes down to "Who do we need to be present as to realize ethical diversity?"

One of the implications of seeing the human–technology relationship as one of strong or coconstitutive interdependence is that by cultivating new ways of being present, we open spaces for creating new technological paths and environments. At least for now, this is as true of our relationship with intelligent technology as with any other previous technology. What makes intelligent technology different is that it is resulting in the presence of artificial and synthetic agencies that are capable of acting at speeds and scales vastly different from those at which we humans can—agencies capable of learning from one another nearly instantaneously how to do better what they have been designed and instructed to do best. This means that the time frame for cultivating more diversity-appreciating and humane ways of being present is not open-ended. We are coevolving with these new forms of agency along lines that seem "destined" to render human intelligence redundant.

If we are going to keep that from happening, we will need to be present in ways that are suited both to resisting the siren calls of digital servants, soldiers, savants, and seers and to redirecting technological dynamics toward realizing systems that support rather than supplant intelligent human practices—systems that foster concentration rather than distraction, that promote commitment rather than convenience, and that are conducive to the effortful cultivation of appreciative and contributory virtuosity rather than the effortless exercise of freedoms of choice.

The New Great Game that is being played by commercial, state, and military actors is premised on individuals acting with rational self-interest in a winner-takes-all competition. These actors are the ones setting and funding global technology agendas and policies. To redirect the Intelligence Revolution, it eventually will be necessary to have corporate, government, and military leaders who are different kinds of persons than has been typical over the last several generations. In Buddhist terms, we will need business bodhisattvas, bureaucratic bodhisattvas, and martial bodhisattvas.

Fortunately, the agendas and policies established by current and future global elites cannot be realized without the compliance of the global majority. The intelligence industries, for the near future at least, can thrive only with our attention and data input. Our attention, however, can be directed elsewhere. To do so consistently, and in the ways necessary to foster ethical intelligence and diversity, we will need to take the lead in becoming new kinds of persons.

6

Dimensions of Ethical Agency: Confucian Conduct, Socratic Reasoning, and Buddhist Consciousness

The core premises of an ethical ecosystem approach to global predicament resolution are that every ethical system brings into focus a distinctive set of actionable possibilities for enhancing relational dynamics; that ethical differences can be engaged as resources for expanding the horizons of ethical consideration and intervention; and that if these differences are marshaled in coordinated response to shared concerns, this will heighten ethical intelligence (adaptive conduct) and establish conditions for the emergence of ethical diversity. For this approach to be viable, ethical deliberation cannot be carried out as a finite game, the winner of which is empowered to set universal standards. It must be engaged in as an infinite game through which the conservative and creative strengths that are constitutive of ethical resilience are continuously enhanced. To realize the positive promise of intelligent technology and change course sufficiently to navigate around or through the ethical singularity ahead will require changing who we are present as to be able to deliberate together with ethical virtuosity.

The aim of this chapter is to place Confucian, Socratic, and Buddhist ideals of virtuosic ethical agency into conversation. In the "Introduction," this turn to classical philosophy was anticipated and justified in part as a route to critical sanctuary: a place from "before" the modern bifurcation of the individual/personal and the collective/societal and from before a human-technology-world relationship that is structured around the values of control, convenience, and choice and energized by algorithmically guided oscillations of human attention between digitally mediated options for autonomy and community. But it is also a route to places "before" the compliance implicit in *being ethical* overshadowed the effortful creativity of *becoming ethical*: places from which to explore the meanings-of and means-to practicing ethical intelligence.

In most of the work being done on the ethics of artificial intelligence, machine learning, and big data, emphasis is placed on establishing behavioral baselines—boundaries defining acceptable and unacceptable conduct in human–machine interactions, at the design stage as well as at the deployment stage. While the resulting principles and guidelines typically include commitments to such universally embraced but relatively abstract human values as fairness and the common good, the primary critical focus is overwhelmingly on such practical and technically tractable issues as privacy protection, transparency, accountability, and explainability. This practical focus is entirely sensible. To have any real-world impact, bridges need to be constructed between abstract ethical values and their technical implementation (Hagendorff 2020).

But, to date, building these ethical bridges has generally involved merely *stipulating* core values rather *struggling* collaboratively with the very deep differences that obtain—across culture and nations—in how values like justice, social cohesion, and freedom are understood and embodied. In effect, while these bridges are solidly grounded on the technical side, they are moveably or flexibly anchored on the ethical side. There are advantages in this arrangement. When the issues being addressed and/or the consequences of value interpretation remain relatively local (national or regional), flexible anchoring can promote ethical variation and experimentation. But when the issues and consequences are global, the effectiveness of ethical guidelines tends to be negatively correlated with interpretative flexibility. What is needed then is not ethical variety but ethical diversity.

The teachings of Confucius, Socrates, and the Buddha were developed in response to questions about who we should be present as in order to realize the conditions of truly humane flourishing. These were not abstract questions. They were questions posed and answered in the context of struggles to develop shared values in the major cultural ecotones of the ancient world. One of the struggles that come with life in a cultural ecotone is the need to find one's way in multiple, overlapping, and sometimes interfusing linguistic environments—a need to reflect on the nature of language and especially on the meaning of words referring to abstract concepts like community and integrity. Confucius, Socrates, and the Buddha were each deeply sensitive to this need.

But the experiential range of cultural difference is not limited to the formidable difficulties of translation or making verbal sense to one another. It also involves reaching accord on what it makes sense to do together. And, in this process, explicit agreements hammered out in words may be much less important than tacit strategies of coordination that emerge out of embodied efforts to resolve relational tensions. Language is not the only—and not necessarily always the best—medium in which to resolve the "predicament of culture": the dislocating experience of being *in* a culture and being compelled to look critically *at* it as if from the outside (Clifford

1988). Experiencing cultural difference involves the intrusion of a meaning-confounding gap between "me" and "my social circumstances"—a space in which what has been simple commonsense has ceased to make sense. This feeling of disjunction or being out of place is often most powerfully felt by the gesturing and expressive body. And just as it's possible to "think" with your hands while woodworking or playing guitar, it's possible "know" in your gut that you are out of joint. Resolving the "predicament of culture" can be accomplished not only at different relational registers in the medium of words but also in the medium of bodily action, and Confucius, Socrates, and the Buddha were all profoundly committed to embodying the meanings-of and means-to predicament resolution.

Looking back to Confucian, Socratic, and Buddhist traditions is a way of restoring to visibility ideals of predicament-resolving personal presence that might otherwise be difficult to imagine from our current technological and cultural vantages. Rather than forwarding formal philosophical systems, each articulates what Pierre Hadot (1995) has termed a distinctive "way of life" (*manière de vivre*)—a Confucian way of life aimed at achieving harmonious community, a Socratic way of life focused on realizing rational autonomy, and a Buddhist way of life focused on cultivating attentive and responsive virtuosity. In each of these ways of life, it is *becoming*—not *being*—ethical that is of primary concern. Seen "ecologically," they afford a three-dimensional perspective on achieving predicament-resolving personal presence.

The Ideal of Confucian Personhood: Resolutely Harmonious Conduct

If we were to invent a traditional Chinese counterpart to the familiar and foundational Cartesian claim that "I think, therefore I am," it would be something like "as we relate, so we are." What matters most in leading good lives is not what we think or know but how we act and with what kinds of sentiment. Working out how to lead good lives is not something best accomplished in our heads but rather in the midst of our day-to-day conduct with and alongside others. For Confucius, this was not primarily intellectual work. It was the social labor of giving birth to humane relationships.

Confucianism is a twenty-five-hundred-year-old tradition of intergenerational transmissions and rearticulations of role- and relationship-qualifying values. As will be true of Socratic and Buddhist traditions, all that can be offered here is a "taste" of Confucianism's distinctive explications of what it means to be both ethically and creatively present. Perhaps the most concise exposition of the Confucian ideal of socially embodied practice is presented in the *Daxue* or Great Learning, a text that came to be revered—along with the *Zhongyong* (The Centrality of the Ordinary), the *Lunyu* (Analects of Confucius) and the *Mengzi* (Mencius)—as one of the canonical articulations of Confucian thought and practice. Attributed to Confucius (551–479 BCE), this concise introduction to the Confucian "way of life" begins with a statement about the importance of effortful self-transformation: "From the emperor to the common people, in all things, personal cultivation is the root." The term translated here as "personal cultivation" is a composite of two characters: 修 (*xiu*), meaning "cultivate," "regulate," "repair," and "reform" and 身 (*shen*), meaning "lived body," "oneself," or "I/me"—a character that in its earliest form depicts a pregnant woman. The implication is that self-cultivation (*xiushen* 修身) is not work on oneself as an individual *human being* but rather as a socially embedded *human becoming*.

The *Daxue* then directly affirms the *social* purpose of cultivating oneself. As exemplified by the ancient sages, the ideal of human becoming is to bring order to the state by first engaging in personal cultivation to set one's one family in order. In cultivating oneself personally, however, the first step is to discern what is apt or fitting in one's own heart-mind. This is done by developing a sincere intention to engage vigorously and virtuously (*de* 德) in the practice of extending knowledge through personally investigating things.

This effort to root personal cultivation in extending knowledge or wisdom (*zhizhi* 致知) through a process of "investigating things" (*gewu* 格物) might be interpreted as call for empirical or scientific study. But that would be to impose a fundamentally foreign matrix of interpretation on the text. The term translated here as "extending" (*zhi* 致) could also be rendered as "bringing about" or "conveying." The term translated as "knowledge" or "wisdom" (*zhi* 知) is perhaps more aptly rendered as an embodied "realization." The extension of knowledge or wisdom is, thus, less a matter of adding more data to a given body of knowledge than it is a process of expanding one's public embodiment or "making real" of the intention to live virtuously.

Turning to the recommended method of realizing one's intention by "investigating things," a more nuanced translation would make evident that while *ge* 格 can mean "to research," it has the more basic connotations of "correcting," "adjusting," and "influencing." The term *wu* 物 does mean "things" in general but also "living beings" and "matters" or "states of affairs." Engaging in *gewu* is thus not just studying things in an objective "hands off" manner; it is *affecting* things as needed to *true* or better align relationships. This sense of the term is brought out by traditional Chinese commentators, who generally held that *gewu* consists in discerning the dynamic interrelatedness of things (*dao* 道)—a corrective restoration of and appreciative attentiveness to the natural order of things.

The root of Confucian personal cultivation, then, is not the discovery of knowledge about what things are; it is making useful distinctions regarding the organically informed coherence (*li* 理) that obtains among them. And since discerning what is apt or useful involves attuning our own heartminds to our circumstances as persons-in-community, working out from within our most intimate roles and relationships in the family, personal cultivation consists in *acts of differentiation* that *optimize mutual contribution.*

In the Confucian world, as in ours, our basic social context is the family. We do not enter the world as generic, individual human beings but as *this* son or daughter in *this* family as part of *this* community. Put another way, our basic human nature (*renxing* 人性) is relational, and personal cultivation is necessarily a process of *qualitative* growth in our roles and relationships. Thus, the *Zhongyong*, a profoundly influential text attributed to Confucius's grandson, Zisizi (483–402 BCE), states that "consummate personhood (*ren* 仁) means comporting oneself in such a way that devotion to one's kin is most important" (ZY 20). And as the *Mengzi* makes clear, qualitatively transforming our relationships is possible for everyone because the human heartmind contains incipient forms or "inklings" of humaneness (*ren* 仁), appropriateness (*yi* 義), ritual propriety (*li* 禮), and wisdom (*zhi* 知), which are rooted, respectively, in feelings of pity for others' suffering, shame at crudeness, modest deference, and approval/disapproval (M 6A6). In the Confucian "way of life," human nature consists ultimately in this potential for qualitative relational transformation.

Thus, when asked about the difference between humans and animals, the *Mencius* allows that the difference is initially infinitesimal. Like animals, human beings eat, procreate, and communicate. But human beings take the act of eating to refuel the body and extend it aesthetically and socially to realize culinary arts and hospitality as entirely new kinds of relational domains. Humans transform the procreative act into one of "making love" and extend that passion for warm connection to generate family relations and community. Humans take animal grunts and cries of warning and pain and through continuous refinement give birth to poetry and song and intergenerational narratives of cosmic scope. Human nature, in a Confucian world, consists in a distinctive propensity (*xing* 性) for relational appreciation, where "appreciation" consists in both sympathetic enjoyment and the generation of value. Confucian community is thus structured around valorizing the personal exemplification of ritually defined roles and relationships (*li* 禮)—roles and relationships into which we were born or assigned by custom, but that we can personalize with greater or lesser vitality and quality.

Undertaken in Confucian terms, then the interrogation of who we are and should be is not a matter of abstract or metaphysical speculation.

Instead, it is a process of discerning how to more deeply and aptly attune ourselves—including emotionally and aesthetically—to our roles as sons and daughters and then as members of extended families and clans, as friends, and as community members. In keeping with the central tenet of the unity of knowledge and action (*zhixing heyi* 知行合一) in Wang Yangming's (1472–1529) revitalization of *daoxue* or the cultivation of the (Confucian) way, knowing who we are is not a result or function of disengaged rational contemplation but of engaging wholeheartedly in socially validated practices that are rooted, finally, in family reverence (*xiao* 孝) and younger-to-elder sibling deference (*ti* 弟).

Confucian persons are not modern individuals. There is no assumption of all persons being equal or of being both naturally and rationally self-interested. Confucian persons are irreducibly relational. And since our most primordial roles and relationships are familial, affirming who we are in Confucian terms is an affirmation of hierarchy—an affirmation of dependence that ideally becomes increasingly mutual over time, but that is never transmuted into anything like liberal independence or autonomy. As the Confucian *Analects* makes clear, even "in seeking to establish themselves, consummate persons establish others; and in looking to promote themselves, seek how to promote others" (Analects 6: 30).

The Confucian acceptance of lived hierarchy as a constitutive fact of personhood in community with others can be unsettling. If unchecked by countervailing values, the social validation of hierarchy can easily result in perniciously self-perpetuating power structures. But hierarchies need not be pernicious if their affirmation is validated by appeals to the value of difference as the basis of mutual contribution. If each of us were equal in every way, we would have nothing of value to offer one another—nothing to contribute qualitatively in the shared practice of nurturing life (*yang sheng* 養生). Total equality may be quite appealing as a social ideal. But as social reality, total equality would be equivalent to maximum entropy: the complete absence of the kinds of differences needed for meaningful interaction to be possible. Hence, the Confucian ideal is one of conserving social differences in pursuit of "harmony, not sameness" (*he er butong* 和而不同), where harmony is not acting in unison but rather adjusting to one another as needed to enjoy the mutual benefits of differing from and for each other.[1]

This, of course, is a Confucian ideal. In actual practice, it often proved to be compatible with profoundly conservative patriarchal institutions and family-centered networks of favor-mediated, mutual loyalty (*guanxi* 關係). As might be expected, it is a perspective on personhood and community that did not go unchallenged. In the cultural ferment and power struggles prior to the unification of China in the third century BCE, other philosophical traditions—most notably Legalist, Mohist, and Daoist—variously expressed concerns about relying on human feeling (*renqing* 人情) rather than law as a guarantor of social order; about the effects of identifying authority with age and tradition rather than practical excellence; about how placing family first can legitimize corruption; about how cultural conservatism restricts creativity; and about the parochialism of overemphasizing the human in relation to the celestial and natural. But these critical responses notwithstanding, from the Han dynasty (206 BCE–220 CE) onward, the warp and weft of the fabric of Chinese society—however much it has been embroidered with principles or coherences (*li* 理) drawn from other traditions, including Daoism, Buddhism, and (most recently) scientific materialism—has been woven in Confucian thread.

The fact that this conception of personhood has been as resilient as it has in the face of often virulent critique warrants taking seriously the possibility that it could contribute valuably to contemporary deliberations about who we need to become to initiate and carry through with global predicament resolution. Indeed, in the context of such deliberations, a distinctive value of a Confucian understanding of harmony-seeking, role-inhabiting persons may be its striking difference from the modern default conception of persons as freedom-securing, rights-bearing individuals acting rationally and naturally in their own self-interest—a way of understanding personhood that has arguably been complicit both in realizing conditions for the emergence of global predicaments and in ensuring their apparent intractability (see, e.g., Rosemont and Ames 2016). Indeed, Confucianism is undergoing a state-sanctioned revival in China today as an indigenous bulwark against the excesses of market-induced individualism, against the narrowness of "egological" rather than ecological thinking, and against the erosion of commitments to intergenerational care.

Confucian traditions offer a powerful vision of humane becoming rooted in the effortful personal cultivation of sensitivities to relational quality. Using family relations as a point of metaphorical departure for understanding the complex patterns of interdependence within and between the natural and social worlds, its vision is explicitly one of establishing improvement conditions for the relational dynamics in which we find and express ourselves. Implicitly, it is a forceful counter to imposing on ourselves and others the success conditions of rationally derived and rationally fixed ideals.

This is *not* a collectivist vision of the suppression of individual creativity. The opposition of the collective and the individual belongs, conceptually, to the modern West. The Confucian vision contrasts, instead, small people (*xiaoren* 小人) and cultivated persons (*junzi* 君子) or sages (*shengren* 聖人)—a distinction between people whose moral and cultural compass or horizons are narrowly constrained and those who can rightly serve as models of appropriate conduct. It is the *junzi* and *shengren* that stand out in Confucian society as personifications of exemplary human becoming. The liability of such a vision, of course, is giving tradition the wrong "face," inviting inappropriate "neighbors" into our midst, and invoking as exemplars people who are out of alignment with real-world relational dynamics.

The Ideal of Socratic Personhood: Critical Resolve

Given this liability, the Socratic citizen is an interesting counterpart to the Confucian sage. Like Confucius and the Buddha, Socrates (469–399 BCE) did not commit any of his own thoughts to writing. What we know of his philosophical "way of life" comes only from secondary sources—most influentially through the dialogues crafted by his student, Plato (427–347 BCE), one of the "founding fathers" of European philosophy. From what we can glean from these sources, Socrates was not a systematic thinker or someone who claimed either great expertise in any particular field of inquiry or in any class of practical endeavor. Instead, his defining trait seems to have been an unflagging desire for or love of wisdom as the key to human well-being.

Socrates's method of acting on this love of wisdom was dialogic. If accounts of his philosophical life are taken at face value, he was passionate about drawing his fellow citizens into tirelessly inquisitive discussions regarding their beliefs and life choices, initially focusing on people who claimed special expertise (in politics, in poetry, and in craft activities), and eventually engaging anyone he happened to meet as he made his way around Athens. By the time he was in his mid-40s, Socrates's cross-examinations of his fellow Athenians apparently became sufficiently pointed to begin leaving his interlocutors puzzled and angry, first earning him a reputation as a public nuisance and eventually leading to being charged with corrupting the youth and religious impropriety—crimes for which he was convicted and sentenced to death at the age of 70 by a jury of peers.[2]

Socrates clearly was not doing philosophy in a harmony-seeking, Confucian key. The now standard arrangement of the Confucian *Analects* opens with a remark about the joys of studying and applying what one has learned, an identification of personal refinement with having no sense of frustration if public acclaim proves elusive, and a statement that it is unheard of for anyone who practices the Confucian "way of life" to defy authority—literally, to "clash with those above" (*fanshang* 犯上) (Analects 1.1–1.2). The picture of Socrates that emerges from the Platonic dialogues and other contemporary sources is of a man who was not only ready to clash with those above and break with societal conventions in the pursuit of wisdom but one who took that pursuit seriously enough to accept death by poisoning rather to disavow his interrogative method. Indeed to the extent that we accept that his trial, defense, and conviction are accurately portrayed in Plato's *Apology*, Socrates apparently used his trial as a high profile occasion to argue, without regard for how it might affect his prospects of avoiding the death penalty, that the highest good was to lead philosophical lives, investigating the meaning of human excellence and all that impacted the wellbeing of both individuals and society. The unexamined life was simply not worth living (see, e.g., *Apology* 38a2–6).

Whereas Socrates's most notable philosophical heirs—Plato and his brilliant student, Aristotle (384–322 BCE)—would develop complex metaphysical and epistemological systems, Socrates himself seems to have been most interested in addressing the "predicament of culture"—that disturbing sense of uncertainty regarding who we are and what we should be doing that occurs when our commonsense intuitions about how things work are apparently no longer reliable. In terms of commitment to interrogating their own actions and values, Confucius and Socrates were close philosophical kin. But Confucius described himself as a transmitter of cultural wisdom, not as an innovator or improviser (Analects 7.1). He sought to resolve the predicament of culture from within his inherited culture by embodying its core values and by personalizing what he came to regard as an exemplary relational grammar (*li* 禮)—a set of constitutive roles through which to articulate the meaning of personal poise and social harmony. Confucian social creativity focused on achieving excellence in the performance of "standard" roles and relationships. Socrates seems by middle age to have lost faith in seeking guidance from the social and political grammars of Athenian life or from the purported wisdom of those hailed as its exemplars. For Socrates, social creativity had to begin at least with exposing the intellectual laziness involved in accepting customary standards and the cowardice involved in choosing complacence over excellence.

The reasons for Socrates's withdrawal of faith in tradition are not known. But it seems plausible that they were existential, not intellectual, in origin. Although Socrates is reputed to have frequented various Athenian gymnasiums—schools in which young men engaged in physical and intellectual training—he was not himself a scholar by training or inclination. He was a soldier. In each of his three military deployments over the course of his adult life, including one lasting three years, he fought courageously and well enough as a foot soldier to be held up as a model of martial valor. These deployments each ended in defeat, however, and he returned from the last of them to find that approximately a fourth of the Athenian population had fallen to plague. Over the final quarter century of his life, he witnessed a steady erosion of Athenian power, increasing internecine conflict, Athens's defeat by Sparta in 404 BCE, and then a short but vicious period of oligarchic rule until democracy was restored not long before his trial and death.

It is not hard to imagine Socrates's battle-hardened eyes taking an increasingly skeptical slant as the knowledge and expertise claimed by those debating and voting on legislation and policy in Athens failed repeatedly to secure the conditions of well-being for all. When considered in light of his cutting and uncompromising conversations with Athenian cultural and political leaders, his refusal to escape trial by fleeing the city-state with the help of friends suggests that what bothered Socrates most were not the failings of Athens's social, political, or religious institutions but rather the personal, moral failings of those falsely claiming the expertise needed to ensure that those institutions would bring benefit to all. What emerges in these conversations is an overarching concern for personifying the quest for excellence (*arête*) and an unwavering faith in the use of reason as the most effective tool for cutting through the personal and social barriers to first beginning and then sustaining that quest.

For Socrates's contemporaries, *arête* had the general connotation of a disposition toward excellence in actively pursuing one's aims. Although it is now often translated as "virtue" to highlight the moral connotations that the term acquired in philosophical circles, *arête* could also be translated as "valor." It was, in any case, presumed to be something directly observable. *Arête* was something one could see exemplified or personified. Socrates seems to have disagreed. In conversing with his fellow Athenians, he was apparently intent on proving, instead, that whatever virtue might be, it was not something immediately evident—something that could be pointed out by example. Over the course of his decades of effort, he failed to find even a single person whose claims of virtue or wisdom stood up to reasoned cross-examination. In modern vernacular, when it came to virtue, he found no one who really knew what they were talking about. None had irrefutable views about any of the major virtues celebrated in Athenian society: courage (*andreia*); reverence or religious piety (*eusebeia, to hosion*); a sense of aptness and prudent moderation (*sophrosune*); uprightness in the sense of doing what just or right in any given situation (*dikaiosune*); and, most importantly for Socrates, wisdom (*sophia*) as a kind of knowing that goes beyond *techne* or skill in getting things done in already familiar ways.

As someone with extensive battlefield experience, Socrates did not question the importance of living with integrity and valor (*arête*). His questions were directed, instead, into the gap between people's claims of knowledge about living virtuously and their actual conduct. Confronted with this gap, his conversation partners were often inclined first to admit their shortcomings in putting their knowledge into action and then to place the blame for these shortcomings on excessively strong passions or insufficiently strong willpower. But Socrates found it unthinkable that anyone who truly knows what is right and just, or what will bring pleasure and happiness, could instead do either what is wrong and unjust or what would bring pain and unhappiness. Although it is still a matter of debate, Socrates seems to have been arguing that wisdom—the virtue included within all of the other virtues—inclines us ineluctably toward what is good and just. The source of our personal and societal failures to do what is good, or to ensure happiness for all, is a lack of real knowledge or wisdom. In this, Socrates is a philosophical kin of Confucian Wang Yangming.

Rather than turning to past exemplars for guidance, however, Socrates turned to reasoning itself. As evidenced in Plato's dialogues, most clearly perhaps in the *Meno*, the decades that Socrates spent interrogating his fellow citizens apparently led him to conclude that while virtue was celebrated by everyone, it had been truly exemplified by none, not even himself. As I interpret this claim, the conclusion Socrates arrived at was that virtue is not a fixed goal or destination—something that we can clearly define or circumscribe. While his philosophical heirs seem to have been inclined to seek the essence of things (*ousia*) hidden behind (or within or above) things as they appear to us, Socrates accepted the pragmatic truth that it is precisely *seeking* wisdom that is most virtuous because all virtues ultimately consist in doing better in all the personal and social pursuits in which we are already truly doing our best. Virtue or excellence is not a *destination* at which we might one day arrive; it is a *direction* of self-transformation.

Realizing the Socratic personal ideal thus entails a forward-looking disenchantment with both past and current conceptions of who we are and should be. This implies, as a matter of philosophical necessity, a willingness to engage in what amount to countercultural uses of reason. The Socratic ideal is one of personally cutting through customary and habitual thinking to transcend current conceptions of excellence, acting critically from within one's culture to further the ever-emerging values of that culture.

The Socratic practice of seeking greater clarity through rigorously rational and passionately habit-challenging engagement with others might be seen as consonant with the modern conception of persons as autonomous, freely choosing individuals engaged in rationally pursuing their own self-interest. But that would be a misreading, I think, of Socrates's commitments to self-transformation as the means to social flourishing. The valorous pursuit of excellence is not undertaken for one's own individual benefit but for the benefit of society and the advancement of its most critically tempered values. Freedom, for Socrates, was not a matter of simply exercising choices. The unlimited choices for information and experience made available by digital media today would likely have held very little appeal for him. Consistent with his career as an infantry soldier, Socrates identified freedom, not with choice but with conducting oneself excellently, even when compelled to carry out duties and actions which one would much rather avoid. His acceptance of the death sentence delivered by his peers was not an expression of resignation or an admission of wrongdoing. It was an expression of valorous commitment to shaking society off its customary foundations and pursuing the well-being of one and all through reasoning freely about the meaning of and means to personal excellence.

Blending Relational Intimacy and Rational Integrity

The Confucian and Socratic ways of life express two very different approaches to addressing the feeling of displacement that occurs when circumstances conspire to render us uncertain about who we are and what we should be doing. At the risk of caricature, we might describe the Confucian method as one of seeking the conditions of culturally sanctioned and relationally articulated *community* and the Socratic method as one of securing the conditions of rationally sanctioned and personally articulated *autonomy*. Yet, both are methods formulated in the midst of socially engaged lives of face-to-face encounters. They are not "armchair" approaches to the philosophical life.

They are also in significant agreement that closing the interrogative gap opened by predicaments of culture is not a matter of discovering or defining, once-and-for-all, the essence of human being. Rather, closing that gap involves assuming new kinds of responsibility for the process of our *humane* becoming. In short, they are in agreement that our journeys of personal self-cultivation are ultimately always interpersonal journeys into a qualitative unknown, carried out in full cognizance of whence we have come. They look to the past, not in longing for it but in order to ensure that they do not unintentionally return to it. Although differently so, they are both progressives in their visions of ethical virtuosity. Both affirm the necessity of personal practice and a readiness to break free from the gravitational fields of the habitual and the customary to engage responsively and responsibly in establishing new life headings.

With respect to the work of shared predicament resolution, there is an interesting complementarity to Confucian and Socratic practices of ethical intelligence. The Confucian life is predicated on refining our conduct, attuning ourselves to the quality of the relationships we share with others, and committing to realizing (*zhi*) the depth of human feeling (*renqing*) needed to engage others with both propriety (*yi*) and consummate humanity (*ren*). The Socratic life is predicated on refining cognition, assessing key concepts and values pertaining to human well-being, and seeking the wisdom (*sophia*) and courage (*andreia*) needed to act justly (*dikaiosune*) and with excellence (*arête*). Taken together, they open possibilities for practices blending attention to relational intimacy and attention to rational integrity as equally important aspects of predicament resolution, shedding valuable light, respectively, on the meaning and practice of shared commitment and clarity.

There is great creative potential in blending Confucian concerns about commitment and conduct with Socratic concerns about clarity as complementary efforts in contributing to the effortful realization of more harmonious and just communities.[3] If there is a liability to this blend, it is that the ethical intelligence that is cultivated in practicing the Confucian and Socratic ways of life is "outwardly" oriented in the sense that it develops and is exercised at a purely social "level of abstraction"—a level at which consciousness is assumed rather than investigated and explored and critical attention is directed at acting and thinking rather than at the process of attending itself. This liability can become acute in the contemporary context of the colonization of consciousness and an attention economy stimulated by algorithmic agencies laboring innovatively to accelerate desire-turnover—a world in which intelligent human practices are being both supplemented with and supplanted by smart services.

It is with this liability in mind that I want to explore blending concerns about commitment and conduct with concerns about clarity and cognition in a space of Buddhist ethical engagement with qualities of consciousness and attention training. Distinguishing between outward and inward orientations of ethical intelligence is, of course, simply a useful convention. In a world of strong interdependence, intelligence is ultimately always recursive: the adaptive revision of adaptivity itself. As noted earlier in passing, the ultimate suchness (*tathatā*) of all things may, indeed, be simply openness or emptiness (*śūnyatā*). But the worlds we inhabit and the lives we lead are conventional, and within them it is "apparent" that turning attention on itself reorients ethical intelligence and opens a distinctive dimension of virtuosic resolve. This, I think, is crucial if we are going to mount sufficiently deep resistance to the algorithmic tailoring of experience to be able to resolve the predicament of intelligent technology and redirect the dynamics of the Fourth Industrial Revolution.

Attentive Virtuosity: The Heart of Buddhist Ethical Effort

Buddhist practice has traditionally been regarded as beginning in earnest with the intent to personify predicament-resolving responsive virtuosity. This is referred to as *bodhicitta*—a term that fuses "enlightenment" (*bodhi*) and "thought/mentality" (*citta*). In Mahayana Buddhist traditions, *bodhicitta* came to refer more specifically to the generative intent to realize enlightenment in order to help all sentient beings free themselves from conflict, trouble, and suffering. That is, giving rise to *bodhicitta* marks the point of setting out resolutely, and with loving-kindness and compassion, on the path of becoming a *bodhisattva* or "enlightening being."

Generating this liberating intent is one thing; acting on and sustaining it is another. As the well-known sayings go, "the road to hell is paved with good intentions" and "talk is cheap." It is one thing to set our sights on becoming an enlightening presence and to talk about relinquishing our horizons for relevance (cultivating wisdom or insight into interdependence/emptiness), our horizons of responsibility (cultivating moral clarity or insight into karma), and our horizons of readiness (cultivating attentive mastery or capacities for responsive virtuosity). It is quite another to embody these intentions within our relational circumstances, whatever they may be. How do we operationalize or enact liberating intent?

Given the karmic nature of the blockages between generating liberating intent and achieving liberating conduct, answers to this question have naturally varied. The teachings and practices suitable for someone blocked by laziness are not identical to those appropriate for someone blocked by anger or by arrogance. But the bottom line is that as long as we are tangled in states of mental, emotional, or physical agitation, as long as we are easily distracted and able to pay attention only fleetingly or by force of habit, and as long as we are caught up in obsessive reflection and calculation, we will fail to be present as needed to perceive and alleviate the conditions and causes of conflict, trouble, and suffering. We will not succeed in becoming present as needed to engage in sustained and shared predicament resolution.

Enacting commitment to karmic transformation and liberation from conflict, trouble, and suffering is simply not possible in the absence of calm concentration and clarity. Developing moral clarity, attentive mastery, and wisdom—the so-called three trainings (*tisikkhā*; *triśikṣā*) around which Buddhist practice is organized—are each necessary to realize enlightening relational dynamics. But attentive mastery or *samādhi* is in many ways pivotal.

The Buddha described his enlightening insight into the interdependence of all things as like coming upon a once-glorious city, long forgotten and overgrown by jungle—in this case, a jungle of perspectives, beliefs, desires, attachments, and cognitive, emotional, and bodily habits. Buddhist practice can thus be described metaphorically as jungle clearing: pulling out the roots of ignorance (*avidyā*; *avijjā*), craving (*tṛṣṇā*; *taṇhā*), and aversion/hatred (*dveṣa*) that nourish the entangling growth of both karmic habituations (*saṃskāra*; *saṅkhāra*) and self-affirming conceptual proliferation (*papañca*; *prapañca*). Meditation is basic training for doing so.

As Buddhist traditions evolved, discussions of meditative realization became remarkably varied and detailed. But as one of the three trainings, attentive mastery or *samādhi* continued to be embraced as an ideal of perfecting meditative presence. *Samādhi* literally means "placing together," but in Buddhism, it refers most generally to a dynamically sustained state of deep concentration: a highly collected or undistracted manner of being present/aware. One of the earliest and simplest techniques for cultivating attentive mastery is mindfulness meditation (*satipatthāna*): engaging in direct and sustained attention to the moment-by-moment condition of body-mind-environment. This means, for instance, sitting still and quietly breathing in and out, without thinking, simply attending to what is ongoing, and then recognizing the beginning of concern about the pain emerging in your right knee or an incipient feeling of sadness or anger, acknowledging them as such, refraining from dwelling either on or in them, and returning to just sitting, just breathing. Unlike meditative techniques aimed at generating specific kinds of experiences, Buddhist mindfulness training aims only at experiential immediacy—the remembrance of things present through awareness cleared of biasing judgements about whatever is occurring or "flowing together" in this situation, at this moment. Mindfulness is abiding in thought-free attentive equipoise.

The other two broad categories of meditation practice, calming/settling (*samatha*; *śamatha*) and clear-seeing/discerning (*vipasannā*; *vipaśyanā*), can be understood as complementary extensions and intensifications of mindfulness. These are attention-training practices focused, respectively, on achieving ever more profoundly settled concentration and on realizing ever greater clarity about the relational origins, meanings, and qualities of experience. Although they are sometimes seen as meditative alternatives, calming/settling and clear-seeing/discerning practices are in fact best understood as simultaneous (and not sequential) aspects of a single discipline.

As described by the Chinese Buddhist thinker Zhiyi (538–597 CE), in his massive and influential treatise on meditation (the *Mohe Zhiguan* 摩訶止観), calming concentration (*zhi*, 止) and clarifying discernment (*guan*, 観) are like the two wheels of a cart or the two wings of a bird. One wheel or one wing, by itself, cannot fulfill its function as a wheel or wing. It is only if we are able to apply (*xingyong* 行用) both calming concentration and clarifying discernment in all of our endeavors that we will find ourselves swiftly and naturally "passing through karmic/causal conditions, correcting/according with circumstances" (*liyuan duijing* 歷緣對境) (T46, No. 1911, 100b).

Zhiyi does not explain how calming/settling and clarifying/discerning enable us to naturally accord with our circumstances and make our way spontaneously through our karmic entanglements. But a hint of why meditation might have this liberating effect is given by an anonymous member of the circle of practitioners that gathered around the legendary founder of Chinese Chan Buddhism, Bodhidharma (*c.* 500): the vital energy (*qi*) of those who attain insight through the medium of written words is weak; the *qi* of those attaining insight from their own circumstances and events by never losing mindfulness anywhere is robust. As the Chan teacher Huangbo (d. 850) insisted, without sufficient vital energy, even if your aim is good, your arrow falls short of the mark; even if your intentions are good, they fail to be aptly enacted.

One of the effects of devoting attention energy to meditation is that it interrupts the polluting "flows" (*āsava*; *āsrava*) of self-affirming views, habit formations, sense attachments, clinging desires, and conflict-generating concepts that normally nourish our karma and enable the "jungle" obscuring our own Buddha nature to remain deep and lush. If sustained with sufficient vigor and over sufficient time, meditation contributes to karmic atrophy and eventually a withering away of cognitive, emotional, and behavioral entanglements.

But no less importantly, attentive mastery affords entry into critical sanctuary—a place of flexibly poised awareness from which to cultivate insight into interdependence and karma: insight into how experiential and relational outcomes and opportunities are shaped by the complexion of our values, intentions, and actions. It is from such a place that it becomes possible to fully experience—and improvise ethically in response to—the interventions of algorithmic agents, the dynamics of the attention economy, and the wish-fulfilling allure of the smart systems through which the colonization of consciousness is being conducted.

If it is true that all assaults on human autonomy and community begin with an assault on awareness, and if self-awareness is crucial to self-regulation, then the algorithmic capture and exploitation of attention is an assault on our abilities to plan, to reflect critically on our own conduct, and to determine what best contributes to our own well-being. Confronted with the dynamics of the digitally networked attention economy, the opportunity to determine our own futures depends on asserting our rights to sanctuary—our *right* to be left alone by the artificial agencies intent upon most profitably shaping and reshaping the contours of our desires and actions. Demanding such a right, however, depends first on critical *recognition* of the complex ways in which surveillance capitalism operates, and this requires keen and sustained mindfulness in relation to how our attention is being manipulated. It is only when attention is thus turned critically on itself that truly liberating *resistance* becomes possible. To successfully reorient the Intelligence Revolution, concerns about qualities of community and rational autonomy will have to be blended with commitments to wise and compassionately improvised presence in the face of the karma-amplifying intelligent technology.

7

Humane Becoming: Cultivating Responsive Virtuosity

Describing Buddhist practice metaphorically as a process of jungle-clearing brings it down to earth. It is practice undertaken in the midst of daily life, including its most mundane activities, not in some loftily imagined realm of abstract or ecstatic repose. In the beautifully direct language with which the Japanese Zen master Dōgen (1200–1253) encouraged a monk unhappily assigned to clean the monastery toilets: "When you lose your money in the river, you look in the river. When you set your horse free at the foot of the mountain, you seek it at the foot of the mountain" (*Eihei Kōroku* 8.7). Doing otherwise is like looking on the porch at night for keys you dropped in the yard because the light is better there.

As we embark on responding critically to the impacts of the Intelligence Revolution, Buddhism points us in the direction of attending first to what is being done with and to our attention in our ever more digitally connected lives. But the metaphor also suggests that there will be more to our ethical labor than the hard, down-to-earth work of freeing ourselves from algorithmic attention capture and compulsive, desire-defined freedoms-of-choice. Cleared land is subject to quick erosion. Once the jungle of habitual thoughts, feelings, and impulses has been cleared, what are we to cultivate in its place? What ideals of personal presence does Buddhism offer for cultivating predicament-resolving and diversity-enhancing ethical intelligence and for mounting humane resistance to the siren calls of digital servants, soldiers, savants, and seers?

Two Ideals of Personal Presence

While undertaking solitary, silent meditative retreats has always been well-regarded as an expedient for deepening Buddhist practice, quietist retreat from the world has never been a final personal ideal. Attaining a calmly concentrated and clearly insightful presence in solo retreat at a mountaintop hermitage is a valid practice. But the ultimate test of such solo training is whether this presence can be sustained in a bustling street market or during a web crawl through online shopping platforms teeming with ad-matching virtual agents. Buddhist tradition has acknowledged that one can attain enlightenment and afterward remain aloof from helping others to do so. One can be a "lone buddha" (*paccekabuddha*; *pratyekabuddha*). But this has never been held up as a personal ideal. Although the Buddha's enlightenment occurred while he was practicing *by* himself, at no point was he ever practicing only *for* himself.

For more than twenty-five hundred years, Buddhists have concerned themselves with clarifying what is involved in being present with enlightening intent (*bodhicitta*). For our purposes, this rich history of reflection can be distilled into two distinctive ideals—that of uncompelled presence and that of compassionately engaged presence. Considering them in some detail is useful, not as personal "destinies" we might select but as personal "directives" for marrying wisdom and compassion and for fostering capacities for resisting and then more humanely and equitably reorienting the karma of the Intelligence Revolution.

The Ideal of Uncompelled Presence

In the earliest strata of Buddhist texts and in today's Theravada traditions, the practitioner's ideal is to become an *arahant* (Skt: *arhat*): a "worthy one" who has personally realized *nibbāna* (Skt. *nirvāṇa*) or freedom from the cycle of birth and death (*saṃsāra*). Although *nibbāna* is often interpreted literally as a kind of final departure from the travails of sentient existence, it is a freedom that can be realized in this life, with this very body. Of the Buddha's students, two hundred and sixty-four men and seventy-three women practiced deeply enough to be able to write poems commemorating their full and final emancipation from conflict, trouble, and suffering.[1]

It is significant then that the Buddha steadfastly refrained from positively qualifying the *aim* of Buddhist practice. Using *nibbāna* (Skt: *nirvāṇa*)—which literally means "cooled down" or "blown out"—to characterize the aim of liberation makes good metaphorical sense. The arahant's attainment of *nibbāna* is often correlated with becoming free from mental afflictions or defilements (*kilesa*; *kleśa*) and from craving forms of desire that are often compared to flames clinging to and eventually consuming a burning log. Yet, the refusal to further characterize *nibbāna* suggests that something more is at stake in not assigning any positive conceptual content to the apparent aim of Buddhist practice.

Given that Buddhist practice aims at dissolving the causes and conditions for conflict, trouble, and suffering (*dukkha*; *duḥkha*), it would seem natural to characterize success in doing so as an attainment of its conceptual opposite: happiness (*sukkha*; *suḥkha*). But even though happiness factors into traditional discussions of the sequence of experiences that practitioners can expect as they gain facility in calming and concentrative meditations, it is not used to qualify the culmination of Buddhist practice. Happiness is good, but it is not deemed a cardinal point for orienting our progress on the Buddhist Middle Path. Instead, the most common guidance given for orienting our practice is to focus on exemplifying *kuśala* or virtuosic conduct.

As noted earlier, although *kuśala* is typically translated as "skillful" or "wholesome," it actually functions as a superlative. *Kuśala* conduct is not just getting things done in a practically and morally acceptable manner; it is doing so in ways that are virtuosic. Thus, the ultimate aim of early Buddhist practice is perhaps best understood as becoming someone who is not only freed from all forms of compulsory presence but who in addition is also *superlatively present*.

This is the main subject of the *Sakkapañha Sutta* (DN 21), in which the Buddha is asked why humans always end up embroiled in conflict and hatred even though they have the intention of living in harmony and without strife and seem to have the resources for doing so. As was discussed in Chapter One, the Buddha first leads his interlocutor through a psychological account of the origins conflict, enmity, and social strife as being proximally caused by jealousy and greed, by fixed likes and dislikes, craving forms of desires, tendencies to dwell obsessively on things, and finally by *prapañca* or conceptual proliferation—a process of dividing the world up into ever more finely wrought units and relations among them, producing an ever more tightly woven net of fixed associations and judgments that at once supports and entraps the craving- and conflict-defined self. The secret to ending interpersonal discord is to cut through *prapañca* by continually evaluating our conduct and continuing on courses of actions only if they both decrease *akuśala* outcomes and opportunities and also increase those that are *kuśala*. Resolving conflicts and freeing ourselves from trouble and suffering is accomplished through sustaining conduct that is resolutely superlative and qualitatively enriching.

The root causes of conflict and contention are also explicitly explored in three sections of the *Atthakavagga* or Octets Chapter of the *Sutta Nipāta*, one of the earliest collections of Buddhist teachings to be put in writing.[2] In ways that further qualify who we need to be present as to engage successfully in shared predicament resolution, these short *suttas* or teachings of the Buddha link conflict with the tendency to set up things as "pairs of opposites." This dualistic tendency is manifest in the relatively innocuous acts of perceiving one thing as pleasant and another as unpleasant or in deeming one person desirable and another as not. But it can also take more consequential epistemic form, for example, in convictions like "this is true, all else is false" or in rank orderings of things, ideas, people, and processes as "superior," "equal," or "inferior." In each case, whether explicitly or implicitly, a position of judgment and power is assumed that then divides and holds us apart, both as individual persons and as communities. In short, conflict is rooted in being present as someone determined to assert his or her fundamental difference and to stand both apart from and over others.

It is, of course, sometimes necessary to stand up for ourselves and protest on behalf of our own way of life and the importance of our own values. But this can be done in ways that assert our individual autonomy or in ways that recognize and take personal responsibility for the quality of our interdependence. The "my way or no way" approach is never the only one available. Becoming someone who is freely and superlatively present means no longer being inclined to think dualistically in terms of is and is-not, independence and dependence, or freewill and determinism. It means cutting not only through the entanglements of obsessive thinking but also through the self-validating "fetters" (*saṃyojana*) of aversive emotions, pride, assumed independence, and attachment to fixed and indefeasible concepts of the good.

If ethical virtuosity is an achievement of the practice of ethical intelligence, and if intelligence consists in adaptive conduct and thus in capacities for and commitments to differing both responsively and responsibly from and for others, then the path to living without conflict cannot be one of eliminating differences through exercises of power. Rather, paths to conflict-free interdependence can only be blazed by having the strength to accord with ever-differing situational dynamics, responding without appeal to any fixed views and principles and without any reliance on past knowledge, traditional rites, or set rules and methods.[3] In other words, the process of dissolving the roots of conflict and contestation—even when these occur in the course of improvisation—can be realized only improvisationally.

The arahant ideal, then, is not only one of being *free from* compelled forms of presence. It is also an ideal of being *free to* relate in nondualistic ways that are emotionally and socially transformative. Thus, when the Buddha is asked to characterize those who have realized the freedom that attends faring well on the path of Buddhist practice, he describes them as suffusing their entire situation with the relational qualities of compassion (*karuṇā*), equanimity (*upekkhā*; *upeksā*), loving-kindness (*mettā*; *maitrī*), and joy in the good fortune of others (*muditā*) (see, e.g., *Tevijja Sutta*, Dīgha Nikāya 13). While meditative realizations of key teachings have certainly been understood as important indices of Buddhist attainment, no less important are capacities for exemplifying what it means to be present in community with others as a nexus of positive, diversity-appreciating, and resolutely virtuosic relational transformation.

The Ideal of Compassionately Engaged Presence

In Mahāyāna Buddhist traditions, the arahant's freedom from the cycle of birth and death (*saṃsāra*) is recognized, but it is often portrayed as being in tension with the twin ideal of personally reproducing the Buddha's realization of freedom from conflict, trouble, and suffering while also emulating his commitment to help others do the same. This is the ideal of the *bodhisattva* or "enlightened/enlightening being" who is compassionately dedicated to helping all sentient beings attain liberation from compulsive presence.

For bodhisattvas, freedom from rebirth is deliberately forfeited as a personal goal. Indeed, as depicted in a wide range of Mahāyāna sutras, the bodhisattva path is embarked upon by vowing to remain among those caught up in samsaric turbulence, working in whatever ways needed to help them escape these conditions and realize liberating relational dynamics. Progress on this path is traditionally regarded as a function of the force of one's commitment to help all sentient beings attain enlightenment—a commitment made compassionately, knowing full well that sentient beings and their afflictions are numberless and that the path of helping them attain liberation is immeasurable. It is the profound intensity and boundless scope of this vow that is said to make it possible for bodhisattvas to respond skillfully to others' needs, demonstrating unlimited responsive virtuosity (*upāya*) while engaging each and every situation from the angle best suited to activating its enlightening potential.

Due to the karmic nature of sentient presence, the bodhisattva path is one of deepening dramatic implication. It is a path that requires attending to how conflicting values, intentions, and actions are generating experiential currents and countercurrents and then discerning how best to take advantage of these currents to navigate toward more liberating relational possibilities. Bodhisattva conduct is not a matter of simple, on-the-spot, one-way, problem-solving action. It is a function of continual, horizon-relinquishing, predicament-resolving interaction. In the terms introduced earlier, bodhisattvas are not master innovators; they are exemplary improvisers. Innovating is a process of developing new ways of arriving at already anticipated goals or ends. Improvising is a process of working out from within existing norms and circumstances in the direction of what is both unprecedented and qualitatively enriching. That is, while innovation consists in achieving already anticipated results, improvisation consists in continually expanding horizons of anticipation.

The bodhisattva vow of working toward the enlightenment of all sentient beings thus functions as a soteriological compass, not a soteriological telescope. It establishes a clear *direction*, not a specific *goal*. To make use of James Carse's (1986) distinction between finite and infinite games, bodhisattvas do not improvise ways of drawing others into a "finite game" that is played in order to finish and win. They improvise ways of drawing others ever more fully into an "infinite game" that is played for the purpose of sustaining and continually improving the quality of play. Innovation requires and is rewarded with increasing *power*: a capacity for determining situational outcomes. Improvisation requires and builds *strength*: capacities for opening new opportunities within existing relational dynamics for enhancing the interests and contributions of all involved. The bodhisattva ideal is to be, not powerful, but strong.

Innovation and improvisation both involve creativity. But as suggested earlier, innovation is best understood as "closed" creativity that solves problems in pursuit of conventionally defined goals and interests. Seen karmically, however, thinking, feeling, speaking, and acting bodily in ways that yield desired outcomes will also afford us further opportunities for engaging in similarly motivated and enacted conduct. In effect, problem-solving and innovation allow us to "zero in" on getting what we want.

That is good. But it is the nature of karmic cycles not just to repeat but to intensify. Thus, to get better at getting what we want, we must get better at wanting. Yet, to get better at wanting requires ongoing experiences of want or lack. And so, as we get better at getting what we want, we increasingly discover that we no longer want what we get. This is the irony, for example, of success in playing finite games of desire. Likewise, the related karma of control is that capacities for greater control can only be developed as long as we experience our situation as always being in need of still further control. And similarly, the karma of choice is such that increasing opportunities for having and making choices eventually involves creating conditions (like those in consumption-driven market economies) in which pressures to choose mount to the point that choice becomes compulsory. In sum, as conventionally good as it is to get what we want, to exert control over our lives and circumstances, and to exercise real choices about who or what to be and do, none of these patterns of conduct are ultimately *kuśala* or liberating

Improvisation consists in "open" creativity that generates new aims and interests, as well as new ways of organizing aims and interests. This is the kind of creativity that is needed to resolve predicaments. While it's possible to specify in advance what will count as a solution to a particular problem, bringing into view a "destination" that we can hope to arrive at by opening a practically viable route to it, predicament resolution consists in establishing a new "heading"—a new constellation of values, aims, and interests in alignment with which we can reorganize and better coordinate our conduct.[4] Trying to address predicaments by means of closed creativity will serve, over time, only to intensify the very conditions that have brought them about. The superlative, improvisational presence exemplified by bodhisattvas consists in the personification of both appreciative and contributory virtuosity.

As described by Chan master Mazu Daoyi (709–788), this virtuosity manifests in "realizing the excellent nature of opportunities and dangers, so that one can break through the net of doubts snaring all sentient beings, departing from 'is' and 'is-not' and other such bondages…leaping over quantity and calculation…so that one is without obstruction in whatever one does. Leaving behind no obstructing traces, one's actions are then like phrases written on water" (*Da zangjing* 45.406b). Progress on the bodhisattva path is accomplished by continually relinquishing our horizons of responsibility to be able to act virtuosically on behalf of others.

Doing so, however, is predicated on relinquishing our horizons of relevance—not only letting go of apparent boundaries between what 'does' and 'does not' matter or concern us but also opening ourselves fully to the currents of meaning flowing through and structuring each situation. In Mazu's poetic phrasing, this means realizing that "the world's myriad things are one's own body" and that what we experience as resistance from the world and others is in actuality our own "paralysis," an incapacity resulting from ignorance of the dynamic, internal relatedness of all things. By cultivating—that is, realizing in practice—the embodied wisdom through which this ignorance is dispelled, we at the same time cultivate capacities for *appreciating* more and more fully what each of us (and all other things and beings) can mean both to and for one another. In doing so, we effectively open up valuably new domains of mutual contribution and engagement.

Thus, in the context of the bodhisattva life of embodying responsive virtuosity and relinquishing our horizons of responsibility and relevance, we are naturally inclined to relinquish our horizons of readiness: to let go of the hesitation-generated standpoint or view (*diṭṭhi/dṛṣṭi*) from which choices are made about whether and how to act or not act. According to the Chan tradition, this means realizing such complete and intense responsive immediacy that we are, without any hesitation, always "according with circumstances, responding as needed" (*suishi yingyong*, 遂事應用), demonstrating an utterly flexible "harmony of body and mind that reaches out through all four limbs…benefiting what cannot be benefited, doing what cannot be done" (*Da zangjing* 408b).

Nourishing Virtuosic Presence: The Six *Pāramitās*

These descriptions of what it means to be present as an arahant or a bodhisattva might seem hopelessly utopian: ideals that are personified nowhere in the real world as we know it and thus toward which it is impossible to make any practical progress. Even if we did manage to engage in enough sustained meditation practice to accomplish a significant amount of "jungle-clearing," it seems obvious that bringing our liberating intent (*bodhicitta*) to fruition would require a substantial amount of some additional kind of work. Along the way, what is going to keep our striving to become superlatively present and to exemplify responsive virtuosity on a resolutely Buddhist track? To push the jungle-clearing metaphor: are there "weeds" we should be pulling out as they appear? Exactly which kinds of new growth should we be nourishing?

The traditional Buddhist answer is that we should nourish those "roots of virtuosity" (*kuśalamūla*) that are most closely connected with becoming compassionately present in ways that embody generosity, moral clarity, patient willingness, valiant effort, poised attentiveness, and wisdom. These are the so-called six *pāramitās*: six qualities of personal presence that Buddhist practitioners should resolve to cultivate.

These qualities are often referred to as "perfections," based on reading *pāramitā* as an extension of the Sanskrit word *parama*, which means the "highest" or "most distant" and which has the general connotation of "most excellent." But given the open-ended nature of the *pāramitās*, "perfection" has to be understood as a verbal noun: a process of continually perfecting or rendering more complete. Alternatively, *pāramitā* has been divided into *pāra* or "beyond" and *ita* or "that which goes" and thus read as "that which goes beyond." This better captures the sense that the *pāramitās* involve going beyond current norms or standards of personal conduct in responding to others' needs, as long as we refrain from taking "beyond" as some final destination. In short, the *pāramitās* are aspirational headings.

Reading the list of the *pāramitās*, it's tempting to see them as personal virtues or character traits like those, for example, around which Aristotle developed his *Ethics*. But that association would be misleading. For Aristotle, no action in and of itself can be deemed morally virtuous. Virtue consists in acting on the basis of an inner stability or equilibrium that enables one to consciously, rationally, and consistently act in ways that are situationally appropriate: doing the "right" thing because it is right and not just because one can see some utilitarian advantage in doing so or because it is socially prescribed. Virtue consists in acting well because doing so has become part of one's nature or character. In short, although it is socially situated, Aristotelian virtue is personally—that is, individually—determined.

Buddhist *pāramitās* are relationally verified. They are dimensions of relational virtuosity. Rather than being virtues or character traits that we might possess and cultivate as individuals, they are personal values—that is, distinctive but interrelated "modalities of relational appreciation"—that are made manifest not in our individual behavior but only in our conduct or the shared ways in which we are karmically led together.[5]

Generosity (*dāna*). The primacy of generosity among the six *pāramitās* is karmically significant. The term *dāna* literally means "to offer" or to "distribute." In Buddhism, it has the specific connotation of compassionate offering that generates future wealth. Earlier, mention was made of the ironic karma of getting what one wants. The karma of offering is similarly cyclic or recursive. Yet, counterintuitively, the "perfection" of offering does not result in the steady depletion of our resources or wealth but rather their steady increase. To get better at giving, we have to continuously be presented with both opportunities and resources for offering something that is valued by others. The "perfection" of generosity, far from being an exercise in steadily depleting and perhaps finally exhausting our resources, is a process of becoming progressively enriched through enriching the lives of others. In a reflexively ordered, karmic cosmos, generosity is mutually enriching.

In traditional Buddhist terms, our intentional acts are causal "seeds" (*hetu*) that sprout when environmental conditions (*pratyaya*) are right and that eventually bear apt experiential "fruit" (*phala*). For those familiar with agricultural practices, the point is clear: it is possible to cultivate better tasting and more nourishing fruit through careful cross-pollination and nurture. Likewise, it is possible to transform the "flavor" and "feel" of our lived experience by attentively altering our "seed stock" and "growing conditions." It is possible to "genetically" modify the seeds of our own future experience. But given that sentient presence is enactive or "world-involving," our "karmic DNA" doesn't just specify the generation of a certain body type or even a certain pattern of perceptual content. Rather, it specifies a continuously and recursively emerging configuration of environing events and intentional engagements with them: the emergence of meaningfully evolving lived realities.

The enriching karma of generosity or offering is powerfully illustrated in the Vessantara Jataka, the last of the several hundred narratives that didactically recount the "prior lives" of the Buddha. As the story goes, in the life before being born as Siddhartha Gautama and realizing complete and unsurpassed enlightenment, the Buddha had lived as Prince Vessantara and dedicated himself to perfecting the practice of unreserved generosity. So profound was Vessantara's generosity that he eventually gave away his kingdom, his wealth, and even his children and wife in response to others' needs and desires. As might be expected, given the didactic nature of the narrative, all of these gifts were eventually returned to Vessantara. Not only that, as further reward, his kingdom was showered with a rain of jewels that enabled all of the kingdom's people to flourish in complete material security and that ensured the royal treasury would be inexhaustible no matter how prolific Vessantara was in his acts of generosity.

As the story of the Buddha's penultimate life, the plot of the Vessantara Jataka makes it clear that being generously present is the dramatic precursor to enlightened emancipation from conflict, trouble, and suffering. It is Vessantara's embodiment of the *pāramitā* of offering that sets the karmic stage for finally fulfilling the commitment—sustained by the Buddha-to-be across many lifetimes—to become an enlightening presence. Looking beyond the Vessantara Jataka, a wide range of canonical Buddhist texts affirm three types of karmically enriching generosity: giving material goods to those in need in society at large, giving material goods in support of the community of monks and nuns, and giving the gift of the Dharma or teachings that enable those receiving them to author their own liberation from conflict, trouble, and suffering.

In each case, it is important that gifts are offered freely—without a sense of obligation and without expectations or "strings attached." Even the subtle anticipation of gratitude from the recipient of one's gift alters the karmic complexion of offering. To be effective as Buddhist practice, giving cannot be carried out instrumentally, even as a means to becoming a so-called good or virtuous person. True generosity dissolves the boundaries between us, eliding what is conventionally mine in *sharing* what is ultimately ours. It is through generosity that truly shared lives are realized.

Generosity can, of course, be practiced more or less skillfully. While there is always merit in freely offering help or material support, if that help or support misses the mark, the karmic consequences will be skewed accordingly. A wonderfully rich illustration of this is presented in the Cakkavatti Sīhanāda Sutta (*Dīgha Nikāya* 26) in which the demise of a wheel-turning kingdom—an ideal or utopian polity—is precipitated by a king offering startup capital to a man who had stolen to feed his family. Upon hearing about this generous response, some people reasoned that they might be equally fortunate if they also stole something. At first, the king presumed that the resulting increase of theft was a function of poverty being more widespread than he had thought and he continued to offer generous assistance to those caught stealing. But when it became clear that thefts were being committed by people who were not poor and who were acting out of greed, the king promulgated a new law that thieves would be publicly decapitated. The use of ceremonial swords to inflict grave violence gave some people the idea of killing those from whom they stole as a way of avoiding capture and punishment. With each new attempt to control the population and restore order, things spiraled further into chaos and the lives of the people became increasingly callous, short, and cruel.

This story well illustrates the liability of addressing an ethical predicament as if it were a problem open to innovative, technical solution. Yet, what is most salient in the present context is the fact that while the king's motive of alleviating poverty was laudable, his method of offering business capital was not an effective way of doing so in the prevailing societal context and had the ironic effect of unleashing previously latent tendencies toward greed and selfishness. Regardless of how "virtuous" the king's motives may have been, his independently undertaken actions did not prove to be virtuosic (*kuśala*) either materially or morally. Skillful offering entails moral clarity.

Moral Clarity (*śīla*). In early Buddhist texts, *śīla*, the term translated here as "moral clarity," is not only narrowly associated with monastic rules and regulations but also has the broader connotation of moral discipline. For both lay people and those ordained as monks or nuns, this discipline is usually identified with observing the so-called Five Precepts: refraining from killing; refraining from taking what has not been given; refraining from engaging in sexual misconduct; refraining from false, pointless, or slanderous speech; and refraining from taking intoxicants to the point of heedlessness. Ordination as a monk or nun involves vowing to uphold a much larger number of precepts regarding one's conduct both within the ordained community and in one's broader social interactions.

But, understood more generally, *śīla* consists in actively embodied commitment to realizing *kuśala* conduct and sharing lives characterized by virtuosic responsiveness and the progressive elimination of the causes and conditions of conflict, trouble, and suffering. This cannot be accomplished by acting in accordance with any existing set of societally prescribed obligations. Neither can it be accomplished by imposing one's own moral norms and standards on others. Instead, *śīla* consists in *perceiving* the currents of value- and intention-generated meaning and action that are presently shaping the karmic landscape and then *participating* in the liberating redirection of those currents. The practice of *śīla* is thus a process of moral clarification—an ongoing, situational purification of intent. The "perfection" of moral clarity is not fundamentally a process of transforming *how* we act but *why*.

Conventionally, moral clarification might be construed as a process of pointing out to oneself and others the difference between right and wrong—distinguishing sharply (and perhaps absolutely) between moral and immoral actions. In the context of Buddhist practice, however, cultivating *śīla* ultimately involves us in discerning—moment by moment and from one situation to the next—which relational interventions will be superlative in alleviating conflict, trouble, and suffering and which will not. And so, while it is arguably useful to see *śīla* as the basis of Buddhist ethics, it is not an ethics that can be readily fitted into the Western taxonomy of ethical systems.

Although broadly consequentialist, the ethics toward which *śīla* points is not predicated on utilitarian calculations regarding how to maximize some particular experiential "good"—for instance, happiness or pleasure. Although concerned broadly with issues of character and quality, the ethics implied by commitment to realizing *kuśala* conduct is not focused on personal character or the individual cultivation of certain socially prescribed qualities or virtues. Instead it is an ethics focused on creatively and progressively transforming relational character and quality. And although rooted in compassionate concerns for the well-being of others, the moral clarification in which Buddhist ethics consists is not deontological or aimed at determining what duties are universally binding on all reasoning beings, identifying inviolable laws of moral nature. It is both situationally responsive and evolutionary. *Śīla* points, in other words, in the direction of an ethics that is resolutely empty (*śūnya*) or nonfoundational: an ethics of *relational appreciation* or qualitatively enhancing the value of the actual relationships through which we are present as *this* person, in *this* community, at *this* time.

Patient Willingness (*kṣānti*). To the extent that ethics involves a reasoned determination of how we *should* live together, ethical systems offer what amount to correctives for how, in fact, we are and have been living together. One of the attractions of ethical systems that take individual agents and their actions as foundational is that they warrant taking ourselves to be morally independent. While it is true that leading a more ethical life entails acting in consideration of others, our success in doing so personally does not depend intimately on their moral commitments. We can live ethically even if those around us do not. The Buddhist teachings of interdependence and karma, especially when viewed in connection with *śīla pāramitā*, call the presumption of moral independence into question. And it in this context, I think, that *kṣānti pāramitā* is most aptly understood.

The Sanskrit term *kṣānti* can be translated as "patience" and implies a capacity for "forbearance" or "tolerance." Yet, as it is used in a wide array of Buddhist traditions, *kṣānti* also has the connotation of willingness or willing acceptance, which is the root meaning of the earlier Pali term, *khanti*. Given that the bodhisattva path is one of assisting all sentient beings realize freedom from conflict, trouble, and suffering, the "perfection" of *kṣānti* or patient willingness is crucial. This is neither a finite path that might be traversed in relatively short order nor one on which those in need of help will always be happily willing partners. It is an infinite path along which it will be necessary to collaborate with others who may resist change: a path that can be followed successfully only if one sustains a consummately patient presence. And this cannot be a *waiting* patience—an attitude of quiet acceptance grounded in certainty about one's convictions and about the inevitability of success. Rather, it must be an *activating* patience—an expression of steadfast readiness to act in ways that inspire others' readiness to act as needed to reorient relational dynamics in conditions of persistent uncertainty.

Given the generally conservative nature of social and cultural institutions, it must be acknowledged that the readiness to engage in moral clarification and "unconstrained conduct" are liable to result in having to bear considerable burdens of unfavorable scrutiny. Commitment to the bodhisattva path will, at one point or another, place one at odds with prevailing societal norms and standards. Remaining on the path is impossible unless one is able to tolerate moral friction. Having accepted the ontological primacy of relationality, however, those who have embarked on the bodhisattva path cannot practice tolerance or forbearance as a tacit form of escape—a self-securing retreat from relational friction. Rather, the perfection of *kṣānti* involves patiently and willingly transmuting the "heat" generated by this friction into a source of appreciative and contributory radiance.

Valiant Effort (*vīrya*). Including *vīrya pāramitā* among the dimensions of being ideally present dispels any doubts about whether the bodhisattva life is one of energetic engagement. *Vīrya* is often translated simply as "effort." But *vīrya* shares the same Indo-European root as the English words "vitality" and "valor." The effort required to engage in bodhisattva action is energetic to the point of being heroic. In Buddhist contexts, *vīrya* specifically connotes heroically enthusiastic and vigorously embodied commitment to advancing along the path of becoming an enlightening presence.

Of the stereotypical images of Buddhist practitioners, two are particularly widespread. The first is an image of a monk or nun, seated in meditation, eyes just barely open, with just the hint of a serene smile gracing an otherwise expressionless visage. The second is an image of a man or woman dressed in loose natural clothing, smiling, eyes twinkling, and exuding a softly caring presence. These are not inaccurate images. As representations of Buddhist ideals for how we should be present, they are entirely compatible with characterizations of those faring well on the Buddhist path as exemplifying compassion, equanimity, loving-kindness, and joy in the good fortune of others. But taking them to be comprehensive or exhaustive representations of Buddhist ideals of personal presence would be like focusing attention solely on the blossoms of a flowering tree and ignoring the branches, trunk, and roots without which these flowers could never have bloomed. Once a seed germinates, continued growth depends on intensely vital striving both downward into the earth and upward into the sky—a simultaneous and courageous penetration of dark and bright unknowns by tender, nourishment-seeking shoots. In the context of Buddhist practice, the perfection of *vīrya* consists in vigorously and valiantly advancing in the direction of virtuosic engagement, both inwardly and outwardly.

The analogy with plant germination suggests that once the protective shell of the self has been split and opened, there is no turning back from practice. There is no alternative to integrating ever more deeply into one's situation, whatever it might be, and nourishing one's enlightening intent therein. Further, the analogy suggests that our flourishing as persons-in-community depends on vigorous effort toward sustaining and advancing what, for the sake of convenience, we can regard as both inner/psychological and outer/social growth. Successful Buddhist practice is not focused first on experiential insight and then later on social action. It is an endeavor in which growth either occurs at once in both of these directions and domains or occurs in neither.

The analogy is, however, far from perfect. Plants grow unreflectively, whenever conditions are appropriate, and with little or no fast intelligence. One might say that plant karma is relatively thin, but generally good, especially given their ecological service of contributing to the creation and maintenance of an atmosphere supportive of animal life. We, however, cannot grow either personally or socially in the absence of adaptive reflection, continuously and intelligently reconsidering what we have been doing and why. Growth in these relational domains depends on embodying continuously refined, critical, and ready responses to ever-changing environments of relevance and concern. This evolutionary process of growth is not easy to sustain. Doing so requires a courageous readiness not only to confront our own shortcomings but also to overcome our own predispositions and habits (*vāsanā*) of thought, feeling, and action as we proceed into the unknown. The perfecting of *vīrya* is not only crucial to developing this readiness but also to addressing the relationally debilitating effects of doubt.

Ignorance is a primary cause of *duḥkha*. In Buddhist contexts, ignorance is not simply a lack of factual knowledge; it is a complex, self-generating "failure to see" (*avidyā*) how we might change or grow in ways that will begin alleviating experiences of conflict, trouble, and suffering. To be ignorant in this sense is to be *subject to* relational blockage. The primary purpose of Buddhist practice is to exit our ignorance—to expand our consciousness in the literal sense of responsively and recursively transforming how we are enactively coupled with our personal, social, and natural environments. Doubt makes this impossible.

Doubt is not mere uncertainty or an awareness of the fallibility of our judgment. It is not being unsure, for example, about the accuracy of our knowledge or about which course of action we should take. Doubt is a doubling of the self, a paralyzing division of intent, an aborting of commitment that cannot be countered simply by acquiring more knowledge or by engaging in further reflection. The only antidote to doubt is the cultivation of an unwaveringly vigorous and effortful presence.

Poised Attentiveness (*dhyāna*). Unwavering focus is indispensable in working to change who we are present as. No matter how vigorous our efforts are, if they are haphazardly focused, the results will tend to be both frustratingly shallow and discordant. As my Buddhist teacher, Ji Kwang Dae Poep Sa Nim, has put it: once you break ground to dig a well, you should keep digging; don't start a hole here and then move and start another one over there where the ground seems softer or the shade is appealing. Embodying an unwavering, undistracted, and acutely focused attentive presence is the perfecting of *dhyāna*.

Although *dhyāna* can be translated as "meditation" in the broad sense of being wholly absorbed in thinking about or taking the measure of something, it is used in Buddhism to refer to the process of progressively calming and centering the mind and as a necessary precursor for the realization of wisdom—the realization of practically useful insight into how things have come to be interdependent in the ways that they are. Traditionally, different levels or depths of *dhyāna* have been identified and enumerated, each associated with a distinct sphere of attentive engagement—for example, physical sensations and desires, feelings of anger or ill will, laziness, restlessness, and such still subtler spheres as those of infinite space and consciousness. But as one of the key dimensions of realizing a liberating personal presence, *dhyāna pāramīta* consists in realizing and sustaining clear and unwavering attentiveness—a fully poised presence.

The perfecting of attentive poise encompasses the three basic Buddhist meditation techniques discussed earlier: calming/settling (*śamatha*) or concentration-directed meditation, clear-seeing/discerning (*vipaśyanā*) or insight-directed meditation, and mindfulness or the remembrance-directed practice of attending to whatever is happening in and about us without judgment or bias. Taken together, these techniques can be seen as a system of meditative training directed toward becoming so wholly—that is, undistractedly and undividedly—present that, in seeing, there is only seeing; in hearing, only hearing; in feeling sadness or grief, only sadness or grief.

Dhyāna is becoming present without-self, without internal dialogue, without a perceiving or narrating subject, without any "me" whose attention might be hooked by passing events and dragged this way or that.

But *dhyāna* has also been traditionally described as a sequential process of attention training that begins by becoming aware, first, of the initiation of attention (*vitarka*) and, then, of how attention is sustained (*vicāra*). As this training deepens, sustaining attention modulates qualitatively, becoming at first a physically uplifting and joyously refreshing alertness (*prīti*) and then an expansive sense of blissful mental ease (*suḥkha*). Finally, the process culminates in the attainment of unwavering one-pointedness (*ekagrata*).

This sequential training toward one-pointed attentive poise was also traditionally associated with a parallel process of withdrawing attention energy from inclinations toward laziness, doubt, anger, restlessness, worry, and sensation-seeking desires. Seen from this angle, the perfecting of wholly poised attentiveness is accomplished (almost ironically) in the "field of distraction" (*asamāhitatva*) as the embodied realization of freedom from both aversion and attraction. Perfecting poised attentiveness is enacting readiness for appreciative and contributory virtuosity in the midst of any and all circumstances.

Wisdom (*prajñā*). Wisdom is often thought of as something that develops over time through a kind of sedimentation process—a layering of experientially derived knowledge that eventually results in more elevated or encompassing views of things than those available to people with less experience. In keeping with this conception of wisdom as a cumulative and mature sense of how things work, wisdom is generally seen as an attribute of those middle-aged and older. Yet, wisdom is not a simple function of how much factual knowledge we possess about the world. It is a complex capacity for discerning how best to act in our current circumstances—a way of understanding how things work that is more than the sum of its epistemic parts. Wisdom is not ultimately about *knowing that* one thing or the other is true but rather *knowing whether* to act in a certain way or not and *knowing how* best to do so.

Much of this contemporary, commonsense understanding of wisdom also applies to the Buddhist concept of *prajñā*. But there are significant differences. First, the Buddhist understanding is that wisdom is predicated on the achievement of wholly poised attentiveness, rather than on the accumulation of a certain amount of experience. That is, wisdom depends much less on previously acquired knowledge than it does on realizing certain qualities of attention and presence: freedom, in particular, from the obstructing "flows" (*āsava*; *āsrava*) of self-affirming views, habit formations, sense attachments, and conflict-generating concepts. Wisdom is realized when we are most deeply, clearly, and vitally committed to the emancipatory process of revising relational dynamics from within. Given this, it is entirely fitting that the personification of Buddhist wisdom—referred to both as Mañjuśri ("Gentle Glory") and as Mañjughosa ("Gentle-Voiced")—was traditionally depicted as a virile and "eternally youthful" (*kumārabhūta*) young man, not an age-stooped, white-haired elder.

The conviction that wisdom emerges with the attainment of certain qualities of attentive acuity and responsive immediacy is, in fact, implied by the Sanskrit word *prajñā*, which combines *pra-* (an intensifying prefix) and *jñā*, a word that can be translated simply as "awareness" but that (like the English "gnosis") implies "investigative insight," especially into matters of spiritual significance. Wisdom is not a function of intellectual closure but of relational penetration and opening. Thus, in early Buddhist traditions, wisdom is described as an achievement of calmly and yet intently sustained attention training and as the situationally personified enacting of the Buddha's enlightening insight that all things are dynamically interdependent and without any fixed essence. Wisdom connotes irreversible progress on the path of realizing liberating relational dynamics. And although later (especially Chinese) Buddhist thinkers would heatedly debate whether enlightenment is "sudden" or "gradual," their debates were not about the experiential immediacy with which wisdom-enacting insight occurs, but about whether a long (gradual) training regime was needed to prepare for translating insight into action. Buddhist wisdom, we might say, is a function of progressively and consummately (not merely cumulatively) aware conduct.

Consistent with our contemporary sense that wisdom implies knowing whether and knowing how to act, *prajñā* is in the broadest sense pragmatic. It is a way of being intimate with reality in the ways needed to get things done. The common, contemporary convention is that wisdom consists in exercising good judgment, based on past precedent—a kind of inductive application of experience. In contrast, *prajñā* enables us to undertake the unprecedented, improvisational "work" of engaging in *kuśala* or virtuosic conduct and contributing to the realization of superlative turns of events. In a single word, Buddhist wisdom is liberating.

It is also compassionate. The penetrating insights with which Buddhist wisdom emerges transform who we are present as personally. Our presence, however, is irreducibly relational. Thus, our transformation personally is also necessarily relational transformation. Insight into the interdependence and interpenetration of all things is ultimately an interpersonal realization of caring and transformative mutuality. Buddhist compassion (*karuṇā*) is not simply "caring about" and "caring for" others. Compassion is practice-engendered sharing in the openly creative process of enacting nonduality and realizing liberating/enlightening relational dynamics.

A productive parallel can be drawn between Buddhist compassion and the contemporary view forwarded by Martha Nussbaum that compassion is an intelligent emotion that is "embodied in the structure of just institutions" and that generates "a breadth and depth of ethical vision without which public culture is in danger of being rootless and hollow" (Nussbaum 2001: 403). But there are some salient differences. Compassion, for Nussbaum, arises on the basis of three judgments: (1) that some seriously bad things have happened to someone or some group, (2) that these bad things were not deserved by those affected, and (3) that the person or group that is undeservedly suffering is a significant element in one's own scheme of aims and interests. Challenging the widespread market-liberal assumption that the self-interest of individual agents is the natural and necessary cornerstone of civil society, Nussbaum convincingly argues that as an affirmation of our mutual vulnerability and mutual relevance, compassion is the ultimate foundation upon which morally just societies are built.

Yet, the ways in which Nussbaum cashes out the three judgments undergirding compassion—entraining the alleviation of others' suffering with the pursuit of our own self-interest and with guaranteeing capability rights for the exercise of individual agency by all members of society—makes it clear that her conception of compassion remains within the horizon of conventional ethical concerns about individual ethical agents, actions, and patients. Buddhist compassion is not the syllogistic result of rational judgments about the serious and undeserved suffering of others whose fortunes are deemed to be linked to one's own. Rather, compassion is an intentional result of the practice of seeing ourselves as strongly interdependent and working to discern and dissolve the karmic causes and conditions of our shared conflicts, troubles, and suffering. That is, compassion is not an intelligent emotion in Nussbaum's sense of an emotion that is essentially cognitive. Compassion is the practice of superlative emotional intelligence: a practice of improvising relational resolutions to the karmic predicaments we experience. Compassion is creatively enacted empathy.

Early Buddhist commentaries thus describe the cultivation of compassion as a process that occurs when one's heart is moved by the conflicts, troubles, and suffering being experienced by others, when a firm intention forms to bring an end to these experiences, and when this intent is pervasively extended. With the bodhisattva ideal, this pervasive intent to alleviate conflict, trouble, and suffering becomes explicitly intergenerational. It is this "great compassion" (*mahakaruṇā*), channeled through the vow to work for the liberation of all sentient beings, that energizes the bodhisattva's demonstration of responsive virtuosity (*upāya*; Ch. *fangbian*, 孝便)—the demonstration of relationally manifest capabilities for and commitments to discerning how best to deploy situational resources to bring about predicament-resolving and ultimately liberating relational dynamics.

Personal Presence and Ethical Diversity in the Intelligence Revolution

The personal ideals of the arahant and the bodhisattva, enacted interpersonally through the practice of the six *pāramitās*, offer a vision of who we need to be present as to lead predicament-resolving and liberating lives. The aspiration to enact superlative capacities for and commitments to embodying generosity, moral clarity, patient willingness, valiant effort, poised attentiveness, and wisdom is to be intent upon the openly creative realization of *ethically intelligent* interdependence.

Understood as adaptive conduct, intelligence is always both embedded and shaped in relation to an environment of actionable possibilities. Buddhist personal ideals are based on commitments to practices that effectively expand the horizons of actionable possibility that until now have defined and constrained our intelligence. The most basic of these practices is engage in conduct that affirms the primacy of relationality, subordinating the individual to the relational, both ontologically and ethically. The Buddhist concepts of interdependence, emptiness, interpenetration, and nonduality are at once results of and directional indices for engaging in this basic practice.

The practice of seeing relational dynamics as karmically configured—attending to the coevoluationary patterning of values, intentions, actions, experiential outcomes, and environmentally manifest opportunities—is insurance against the tendency to separate out a knowing subject independent of its known objects. Ultimately, this consists in actively annulling the divorce of reality and responsibility that is implied by affirmations of subject–object independence. The further practices of seeing all things as characterized by conflict, trouble, and suffering, by impermanence, and by being without any fixed essence or self—like the more generally formulated practices of relinquishing our horizons of relevance, responsibility, and readiness—is a system for actively extending the scope of our karmic considerations and, thus, for continually scaling up the reach of our ethical intelligence.

These are practices for becoming present in the ways needed both to appreciate and contribute to ethical diversity: intelligent practices for fostering superlative capacities for and commitments to differing responsively and responsibly from and for others. In many respects, they are practices compatible with Confucian and Socratic strategies for humanely improving the quality of interpersonal community, and they resonate well with Confucian and Socratic emphases, respectively, on embodied mutual attunement and on presupposition-challenging cognitive refinement. The Confucian valorization of culturally sanctioned harmonious community and the Socratic valorization of rationally sanctioned personal autonomy point toward relational intimacy and rational integrity as crucially important aspects of predicament resolution. Taken together, they shed valuable light on the meaning of enacting shared commitment and clarity.

As epitomized in the practice of the six *pāramitās*, however, the Buddhist ideal of personal presence blends the values of behavioral harmony and cognitive autonomy in a space of concerns about qualities of consciousness and attention, adding into the mix an explicit valorization of improvisation or open creativity. This underscores recognition that predicament resolution often depends on conduct that is unprecedented and that may run counter to social expectations and cultural norms. In a world of strong interdependence, intelligent practices must be resolutely recursive to remain relevant and humane. They must be practices that embody ongoing, adaptive revisions of adaptivity itself, continuously expanding what is encompassed by the responsive process of becoming truly and responsibly humane.

The Intelligence Revolution is bringing about entirely new kinds of actionable possibilities in environments that are designed to be as seductive and choice-rich as possible. Commercial, state, and military actors have built the basic infrastructure for an attention-exploiting, information-generating, and intelligence-gathering grid that is capable of powering desire-inducing and wish-fulfilling smart systems that will tirelessly interpret, respond to, and inflect our intentions with inhuman speed and precision. If the analysis undertaken earlier proves to be at all accurate, this infrastructure will enable an unprecedented and unfortunately inhumane extension and intensification of the reach of the competition-driving values of control and choice. It has already proven capable of supporting seemingly contradictory exercises of power, enabling both social connection and terroristic disruption, increasing both predictive accuracy and existential precarity, fostering both greater equality of market access and deepening inequalities of market success, and vastly expanding the range of individually exercised consumer choice while no less vastly extending the reach of corporate and state control.

The Intelligence Revolution is bringing about a transformation that is designed to be experienced as almost instinctually attractive. Computational factories operated by ambient intelligences are suffusing our everyday environments with new conveniences, choices, and modalities of control. They are doing this so deftly and intangibly that they are being incorporated into our daily lives almost without question and with a swiftness and sureness that makes a "smarter" global future seem all but predestined.

To be sure, existential concerns have emerged about the threats posed by an artificial intelligence (AI) singularity. Robotics designs and military applications are now increasingly subject to ethical scrutiny. Worries about job losses and the effects of data mining and eroding informational privacy are no longer academic curiosities and are gaining some real public traction. But given our collective track record of ineffectually acknowledging and addressing the more obviously threatening global phenomena of industrially produced climate change, it is hard not to fear that our emerging critical awareness of a looming intelligence predicament will turn out to be too little, too late. To convincingly imagine otherwise, we will have to substantially reimagine ourselves.

The modern liberal practice of imagining ourselves to be inherently free and self-governing individuals was remarkably effective in breaking down millennia-old constraints on who we could be present as. But the ideal of autonomous individuality is particularly liable to co-optation in the commercial build-out of digital and digitally augmented environments that are algorithmically tailored to be maximally alluring for each and every one of us—environments in which experiential choices will become virtually unlimited and from within which truly shared realities are already beginning to be experienced as unnecessary and undesirable spaces of interpersonal friction. In the context of the global attention economy, seeking individual autonomy as a personal ideal can easily be nudged into identity with a rationally self-interested opting-out of the hard, clarity-enhancing, and commitment-building work of global predicament resolution.

Buddhism offers a relational alternative. The practices associated with embodying the arahant and bodhisattva ideals are, first and foremost, methods for enacting commitments to superlative predicament resolution: methods for realizing liberating relational dynamics from within the dramatic complexity of our karmic entanglements with one another. But they are also a systematic means to manifesting the kind of attentively poised and wisely compassionate presence needed to resist the attractions of the attention economy and to foster the ethical intelligence needed to ensure that the Intelligence Revolution unfolds along more just, diversity-enhancing, and equitable headings. The question begging to be asked, of course, is whether this alternative and its ideals offer any realistic vision of how to proceed socially and politically, not just personally.

8

Course Correction: A Middle Path Through Data Governance

We are moving with exponentially increasing speed on a path of material and moral transformation. Smart cities and smart capitalism are enhancing efficiencies, improving access to services, reducing waste, streamlining maintenance, multiplying choices, and magnifying convenience. The internet of things is enabling intention-anticipating, text- and screen-free connective immediacy. And AI fact-checkers and provenance sleuths are helping to rein in disinformation and deep fakes.

Yet, smart security and governance are enhancing population-shaping as well as population-surveilling powers. The spaces of opportunity opened by the Intelligence Revolution include novel openings for both "dark web" criminal activity and "deep state" political abuse. The network structure of the Attention Economy 2.0 is biased toward monopoly and amplifying global inequalities. And the uses of human attention energy and intelligence by corporate actors to design and deliver increasingly desirable and personalized goods and smart services are already telegraphing potentials for rendering human intelligence superfluous.

There is plenty of room to debate the ratio of promise to peril as societies worldwide are suffused with intelligent technology. Every new smart tool and artificial intelligence (AI) application expands the horizons of what is and can be transformed by intelligent technology. There is not, however, plenty of time for course correction. The most perilous thing about the Intelligence Revolution today is its acceleration.

Intelligent technology is holding up a mirror to humanity. In it, we can see with ever-greater resolution deep conflicts among our own values. The predicament at the heart of the Intelligence Revolution is a human predicament. But the mirror of intelligent technology is endowed with adaptive creativity. It is a mirror with agency that reflects on our pasts, shows us more of what we have wished for, and then delivers what we choose. The faster this cycle plays out, the more profound is our captivation. At some point, the thought of exercising exit rights from this arrangement will no longer cross our minds. We will have forfeited experiential and relational wilderness to take up happy fulltime residence on karmic "cul-de-sacs" crafted in minutely detailed response to our digitally expressed values and desires, enjoying residences wherein we can lead compulsively attractive lives, paid for with the irreplaceable currency of attention—lives in which we will never have to learn from our mistakes or engage in adaptive conduct, freed from the need to exercise our own intelligence.

It took roughly two centuries for the fossil fuel-fed industries of the Second and Third Industrial Revolutions to scale up human intentions to the point that it became apparent that continuing to do so would result in irreversible and profoundly disruptive changes in the Earth's climate. In the fifty or so years since then, humanity has not succeeded in rising to the challenge of generating clear and practically effective commitments to shrinking and eventually replacing the carbon-based economy. The scaling up of human intentions and value conflicts by means of the attention- and data-nourished machine intelligences of the Fourth Industrial Revolution will not take centuries to reach a comparable point of causing irreversible and disruptive changes in the anthrosphere—the sphere of human interactions and potentialities and its varied cultural and social climates. The ethical singularity, the point at which the art of human course correction becomes an historical curiosity, is likely to be upon us in a few decades or less.

Our situation is not intractable. Confidence that this is so is one of the great benefits of adopting a karmic perspective on time. Confidence that we *can* set the Intelligence Revolution on a different heading, however, compels asking *whether* we should do so, *when*, and exactly *how*. One possibility, often explored in science fiction stories, is to put the brakes on hard. Given the ongoing and intimate dance of elite corporate and state actors spearheading the expansion of digital connectivity and smart systems, the prospects for a full halt by legal or regulatory means is at best minimal. But given that the advances of the Attention Economy 2.0 and of the colonization of consciousness are proportional with the amount of attention and data that the global majority spends on the affordances of digital connectivity, it is theoretically possible for us to stunt or stop those advances. This power is a function of the peculiarity of serving simultaneously as *consumers* of individually targeted material and informational goods and smart services, and as *producers* of data without which machine intelligences cannot evolve and improve their effectiveness. If enough of humanity abstains from digital connectivity, the Intelligence Revolution can be put on hold.

Digital Asceticism: A Way Forward?

Unfortunately, abstinence is a very hard sell. As public health and family-planning experts will readily attest, while sexual abstinence is a foolproof way of avoiding implication in a pregnancy, it is notoriously ineffective as a birth control policy, either public or personal. The prospects of long-term technological abstinence are arguably much worse. Sexual abstinence means forfeiting entry into just a single domain of sensual pleasure. Abstinence from digital connectivity and participation in intelligent technology means giving up relational possibilities and ranges of experiential choice that, even a generation ago, were scarcely imaginable. The self-control required to sustain technological abstinence would be hard to overestimate. But even if it were possible, abstinence is not ultimately a *kuśala* option. Digital asceticism is not, by itself, conducive to alleviating or eliminating conflict, trouble, and suffering.

The reasons for this are complex. To begin with, even before the COVID-19 pandemic, steadily increasing numbers of people around the world were working in the so-called gig economy, their livelihoods directly dependent on their own and others' use of the attention/data infrastructure. With global mass experiments in online schooling and work-from-home in emergency response to the pandemic, livelihood dependence on digital connectivity is likely to accelerate. In addition, for the generation born into the world of 24/7 connectivity, technological abstinence would be akin to identity-threatening exposure to social vacuum. Continually updating online profiles has become foundational for personal identity, and digitally mediated exchanges are now crucial validations of social existence (Twenge 2017). It might not be that severing ties to social media and the forms of cultural belonging made available through digital connectivity would amount to social suicide, but it would certainly be experienced (at least initially) as a severe, self-inflicted social disability.

In light of the increasing virtualization of the public sphere and the almost complete and attention-capitalizing digitalization of news, information, and opinion communication, committing to technological abstinence is also tantamount to committing oneself to social impact irrelevance. If it is true that our relationship with technology is not merely interactional—a relationship of weak interdependence—but rather a co-constitutive relationship of strong interdependence, placing ourselves entirely beyond the reach of intelligent technology amounts to placing ourselves out of position for significantly affecting the quality and dynamics of that relationship. The human-technology-world system can only be changed from within.

Finally, there is a strong ethical case to be made against any course of action that would compromise the effectiveness of the global intelligence infrastructure or stand in the way of building robustly "smart societies." More than half of humanity now dwells in urban areas and nearly one in four people now lives in slums,[1] and the collapses of national economies around the world during the COVID-19 crisis have led to a great many more being pressed into lives without even the minimal dignity of clean water and adequate nutrition.

Smart devices deployed in municipal water systems can drastically reduce waste and help ensure sustainable access to safe drinking water. Smart transportation management systems can markedly lower congestion and increase urban livability. The spread of smart electrical grids, smart appliances, and waste heat adsorption air-conditioning systems that are optimized by genetic algorithms and artificial neural networks can vastly lower global energy consumption and carbon emissions while also providing relief from the daily life discomforts of a heating planet. Precision farming and algorithmically tuned food processing, shipping, storage, and marketing can contribute significantly to making food security a reality for all. And there is little doubt that the universal availability of virtual medical diagnosis and prescription services would benefit the half of humanity that currently lacks access to even basic health services.[2]

Acknowledging the promise that intelligent technology has for solving problems that until now have tenaciously caused so much human misery, it is hard to call with an ethically clear conscience for the halt of the Intelligence Revolution. Indeed, we are being confronted with an intelligence *predicament* only because the inequity of the "winner takes all economy," the ominous potential of state surveillance and opinion control, and the madness of an arms race in autonomous weapons systems and influence machines are collectively counterbalanced by the scientifically crafted allure of desire-anticipating and wish-fulfilling net sirens, by the promise of horizonless platforms for social connection, and by the potential to finally secure dignified lives for all as machine and synthetic intelligences find social welfare problem solutions that may have been "hiding in plain sight" for decades amid the vast stores of data that digital connectivity has now made both available and usable.

Digital Hedonism: An Impoverishing Alternative

Yet, if technological abstinence and digital asceticism are not *kuśala* responses, neither is the unconstrained, full-speed-ahead alternative of digital hedonism. As is the case with physically indulged hedonism, indulgence in technological hedonism is conducive to proportionate callousness about the suffering and trouble of others. This need not be the extreme callousness of the kind that characterizes those, for example, who frequent brothels or who insist on luxurious lifestyles made possible by servant or slave labor. Hedonistic indulgence is predicated on *carefree attention* to satisfying one's individual desires—ignoring costs to others and thus turning away from the felt interdependencies crucial to compassionate practice. In the case of indulging freely in the pleasures and powers afforded by digital connectivity, it is predicated (among other things) on turning blind eyes to such structural risks of digital indulgence and the proliferation of adaptive smart services as mass unemployment and underemployment and the poverty resulting from them.

Poverty is often identified with an individual's or family's lack of adequate income, material goods, and resources to meet daily subsistence needs. From a Buddhist perspective, however, poverty consists in blockage-riddled interdependencies: the proliferation of *duḥkha*-generating impediments to relating freely. Poverty is not an affliction of the poor, it is an affliction of broken or distorted interactions among "the wealthy" and "the poor": evidence of denied relevance, responsibility, and readiness and of "carefree" expenditures of attention-energy satisfying one's own cravings for independence in ways that cast long shadows of dependence over others.[3]

The karma of digital hedonism is to seek pleasurable and profitable difference from others without considering how to valuably differ *for* them as well. Neglecting to care about and for others is not merely a personal shortcoming. It is a crucial contributor to structurally generated mass unemployment and underemployment that cannot be offset by compensatory measures like a universal basic income because the poverty resulting from them is not merely a matter of deficient income and property. Poverty is the relational indignity of being forced into a position of having nothing to offer to others that they would value. It is being relationally bereft of possibilities for generosity. Failing hedonistically to differ *for* others denies to the impoverished opportunities for making even the most humble offering of all: that of heartfelt gratitude.

But even setting aside explicitly Buddhist considerations, there are compelling reasons to inhibit the hedonistic, "full speed ahead" embrace of intelligent technology and its intention-interpreting, desire-anticipating, and wish-fulfilling potentials. At a purely practical and technical level, putting the brakes on the Intelligence Revolution is necessary to have any realistic chance of having the time needed to ensure, first and foremost, that the internet of things is secure by design. The lamentable reality is that human wishes and intentions are not all benign. Many are willfully malign, and as Bruce

Schneier (2018) has very persuasively argued, the pace with which everyday consumer products and societal infrastructure are being connected and made "smart" carries tremendous practical risk. Every internet-connected device is a hack-portal into our lives as consumers and as citizens.

Putting the brakes on the Intelligence Revolution is also crucial for gaining the time needed to explore, both ethically and legally, the personal risks generated by the business-as-usual functioning of the data- and intelligence-gathering infrastructure of the Attention Economy 2.0. Digital hedonism—unconstrained indulgence in the pleasures of digital connectivity—is not limited to consumer/user pleasures. It includes commercial pleasures as well, the unconstrained pursuit of corporate profit through data analytics central among them.

Commercial indulgence in the pleasures of free data access have been limited thus far primarily by public concerns about personal data privacy. In keeping with the core technological values of choice, convenience, and control, two approaches to securing data privacy have become mainstream: opt out and opt in. In the former, the default position is that data can be gathered, stored, and shared by the companies whose digital platforms and services we use and we are given an opportunity (perhaps deeply buried) to choose opting out of that default. The latter places up front limits on what platform and service providers can do with our data prior to gaining our expressed consent. For the most part, concerns center on whether information *about* us is publicly accessible, exactly how publicly, and the limits (if any) to which our data property can be capitalized upon by others. Here, we are given opportunities to choose which data rights to delegate to commercial interests in exchange, typically, for expanded or improved user/consumer experiences.

Data: Digital Representation or Digital Extension?

Yet, as we discussed earlier, seen in relational terms, our digital presences are not best understood as representations of us but as extensions. The data resulting from digital connectivity is not only *about* us. In a sense, it *is* us. When it comes to user-accessible social media content, for example, if someone manipulates that content without our express permission, we might well experience that as a kind of assault, a personal violation. But what about ad clicks, video views, search patterns, or other data generated by our participation in digital connectivity? For most of us, this amounts to "data exhaust"—the useless informational residue of our connective behavior. So what if the corporations providing us with "free" platform access and digital services turn a profit by recycling it? After all, we are not being confronted—even in illiberal states like China, Russia, or Iran—with gray, Big Brother–style totalitarianism. We are being attended by handsomely presented and very solicitous "Big Others" who become so keenly attuned to our individual complexions of desire and denial that their algorithmic nudges are nearly indistinguishable from moves we might have intended on our own.

Granted that our data is really part of us, our digital traces are more like strands of hair or flakes of skin we've shed. Analogous to how these traces of our physical presence contain genetic information about our physical stature, our looks, and our health propensities, our digital data contains intimately revealing information about our patterns of values, intentions, and actions. In a sense, our digital data is a "genetic" record of our karmic presence. Crudely accomplished, creating a digital profile of someone and running predictive analytics on it amounts to creating a credible simulation. But, if sophisticated enough, doing so could amount to something much more like karmic "cloning": the digital recreation of an intentional presence. Analyzing and then employing people's data to alter their behavior through precisely timed and directed news feeds, ads, search results, and so on amount to karmic gene editing.

This is not yet being done with the precision that CRISPR techniques allow in manipulating physical genes. But even far short of karmic "cloning," there are ethical and legal complexities in the uses of our data that are akin to the issues raised by the use of human genes, for example, in scientific research. The legal and ethical battles over Henrietta Lacks' immortalized cells—kept alive since her death in 1951 for use in medical and scientific research, some of which have resulted in Nobel prizes and the amassing of commercial fortunes—are still not settled. Battles over the uses of our digital, karmic genes have not even begun in earnest.

A Digital Middle Path: Resistance for Redirection

There are thus two critically important tasks before us: (1) to change who we are present as to be able to deliberate effectively and collaboratively about how best to resolve the value conflicts troubling the human-intelligent-technology-world relationship and (2) to slow the intelligent technology design and deployment juggernaut sufficiently to have time for an ethical ecosystem to emerge out of diversity-enhancing and predicament-resolving intercultural, international, and intergenerational deliberations—an ecosystem supportive of efforts to orient the Intelligence Revolution more humanely through the creativity-enhancing and ethically-strengthening coevolution of human and machine intelligences.

The Buddhist Middle Path offers guidance for both. The Buddhist personal ideals of compassionately wise and virtuosic responsiveness, and the moral and meditative practices of which they are achievements, are a cogent path for establishing and enhancing both capacities for and commitments to shared predicament-resolving ethical improvisational. The explicit regard that Buddhism accords qualities of attention and consciousness, its emphasis on the primacy of relationality, and its karmic ethics of continually raising ethical standards make it particularly fit for the purpose of facilitating critical engagement both with the structural ramifications of the new attention economy and with the recursive complexity of the human-intelligent-technology-world relationship. Especially when blended with Confucian and more broadly communitarian concerns about commitment and conduct, and with Socratic and more broadly liberal concerns about clarity and cognition, Buddhist ethics opens a "middle way" beyond the opposition of ethical universalism and relativism and in the direction of sustainably realized ethical diversity.

The Middle Path also affords guidance regarding the strategic slowing of the Intelligence Revolution. In early Buddhist traditions, the Middle Path was often presented as an alternative to both asceticism (as a means to spiritual self-realization) and hedonism (as a means to sensual self-gratification). Understandably, this has often conjured up associations with Aristotelean appeals to moderation—a situationally determined "golden mean" between virtue and vice. But the term also was used with reference to the spectrum of metaphysical views between asserting, at one extreme, the ultimate reality of spirit and, at the other, the ultimate reality of matter as well as to opposing assertions about freewill and determinism. Rather than a practice of taking up some position midway between opposing extremes—whether moral, metaphysical, or ideological—the Middle Path is more aptly characterized as one of discerning how to move skillfully "oblique" to every spectrum of opposing views.

Thus, in instructing political leaders, the Buddha did not denounce or insist upon any specific institutional arrangement and refrained from labeling any of the prevalent systems—aristocratic, democratic, or monarchic—as good, bad, or indifferent. Instead, his instruction focused on taking up some key aspect of standard political conduct and pressing for a *kuśala* understanding of it. Hence, although one of the basic Buddhist moral precepts is to refrain from taking life, the Buddha did not ask a king planning to sacrifice a horse to improve the welfare of his people to abort the ritual. Rather, he asked instead about the real meaning of "sacrifice" and through this deftly guided the king toward a less karmically fraught method of improving people's lives. Generally stated, the Middle Path strategy is not to *refute* and then *replace* others' views but to gently *resist* errant intentions and *redirect* their attention.

This resistance-with-redirection strategy for karmically significant changes seems particularly apt in response to digital asceticism and digital hedonism. Turning again to issues of data governance, it is apparent today that a wide spectrum of views exists encompassing at one extreme a competitive, bottom-up, market approach depending almost exclusively on corporate self-regulation and consumer choice (as in the United States); in the middle, a consultative, peer-to-peer, social welfare approach that emphasizes a substantial role for state intervention to ensure consistency with liberal values (as in the European Union); and toward the other extreme, a constructive, top-down, state-mandated approach that balances economic growth with social stability (as in China). A Middle Path intervention aimed at resolving the predicament of intelligent technology would not be to refute or outright reject any of these approaches. Neither would it be to adjudicate among or rank them. Instead, it is to identify opportunities intrinsic to each for bringing about *kuśala* changes in data policies and practices.

Because the reach of the intelligence predicament is global, rather than national or regional, a Middle Path approach would necessarily also involve discerning opportunities for convergence among these different streams of policy and practice change. This would be done, not with the goal of ultimately arriving at a common regime for global data governance but with the aim of braiding of these streams in a way that conserves their individual integrity while also coordinating their distinctive contributions to a global data governance ecology in which each enjoys a significant, resilience-strengthening share.

In doing so, considerable responsive virtuosity and ethical creativity will be required to improvise pathways from currently disparate positions regarding, for example, the relative value of privacy and social cohesion. As much or more virtuosity and creativity will be needed to braid together national and corporate commitments to transforming the New Great Game from a finite competition for domination in the colonization of consciousness into an infinite game played to enhance relationally emergent qualities of consciousness, especially in the new domains of sentience that are manifesting virtually as human and machine intelligences coevolve.

The Middle Path strategy of resistance and redirection, unlike oppositional strategies of refutation and replacement, does not come with roadmaps. The considerations that compel taking an ethical ecosystem approach to resolving the intelligence predicament also compel accepting the impossibility of individually drafting a global itinerary for realizing the most humane potentials of the Intelligence Revolution. The clarity and commitment needed to resolve the intelligence predicament will only develop and deepen to the point of being practically effective when the compass of collaborative contributions crosses a still-to-be-discovered threshold of sharing. The emergence of ethical diversity cannot be blueprinted. In a world of machine intelligences adaptively intent on scaling up complexly layered human intentions, resistance will have to be improvised and course changes will need to be constantly adjusted.

Intellectual justifications for improvised resistance and redirection notwithstanding, offering nothing in the way of concrete proposals for "next steps" is not a satisfactory option. Silence may be golden, but gold does not make good garden soil. Those advocating for growing resistance and coordinated redirection are in a position not unlike that of Chan master Linji Yixuan (d. 866), who, when asked by a provincial governor to explain Buddhist enlightenment during a public lecture, looked out over his audience and proclaimed that as soon as he opened his mouth to speak, he had already made a mistake. After a dramatic pause, he then admitted that while this was true, it was just as true that offering those attending the lecture nothing to take hold of would be like handing over a fishing net without a drawstring—a net that will never land any fish, no matter how many times it's thrown into the water and wrestled back ashore. To resolve this dilemma, Linji invited anyone who was willing to place themselves wholly on the line to step forward and enter into "Dharma combat" with him—a live, jointly improvised demonstration of enlightening conduct.

Thinking and talking about responsive virtuosity is not the same as embodying it. Enlightenment is not something down the road that we can achieve *through* Buddhist practice; it is an achievement *of* that practice. Thus, Linji famously demanded that if you "see the Buddha on the road, kill him!" Objectifying enlightenment as a goal to attain is subjecting ourselves to a kind of Zeno's paradox. To get halfway, we first have to go half of halfway and, before that, half of half of halfway, ad infinitum. In our present case, the only viable answer to the question of how to go about resisting and redirecting the dynamics of the Intelligence Revolution is to get up; join the dance among commercial, state, and military actors from whatever angle one can; and set resolutely about eliciting moves that will change the nature of the dance.

Building on our conversation thus far, I want to offer two sets of proposals for changing the way things are changing. Broadly speaking, the first set works out from our dual roles as digital consumers and producers and sketches activist possibilities for resisting current data regimes and reimagining and redirecting data governance in pursuit of global public good. The second set focuses on changing the way education is changing, resisting its soft colonization by Big Tech and pursuing a redirection of education in alignment with diversity and equity as explicitly relational values suited to supporting learning practices that are consistent with both deepening and expanding humane creativity.

Connectivity Strikes: Leveraging Citizen and Consumer Data Strengths

Into the near future at least, the data dependence of machine learning systems and smart services accords everyday global consumers and platform users unprecedented leverage. Data withholding retards algorithmic innovation. And, given that, organized social media and e-commerce strikes, undertaken at sufficiently large scale, can function in ways that are not unlike labor strikes in the brick-and-mortar era of manufacturing and retailing. By temporarily stalling or slowing computational factory production, they can buy time for renegotiating the terms of our personal and social contracts with the global purveyors of smart services and systems.

Over the last five years or so, there has been growing popular consideration of the benefits of engaging in social media fasts—temporary commitments to technological abstinence. In keeping with the emphasis on the individual consumer/user that is fostered by the attention economy and by the purveyors of smart services and 24/7 connectivity, these benefits have been presented as almost entirely personal: better sleep, lower stress, making fewer unhealthy comparisons with others' online profiles, and so on. In short, connectivity fasting has been forwarded as a way of addressing the individually experienced effects of an overly rich diet of social media and internet use. Rightly so, it has not generally been seen as a potent form of critical public action and citizenry rehabilitation.

But while connectivity fasting might be relatively powerless in terms of bringing about political or socioeconomic change, the same cannot be said for the collective intentionality and action involved in *connectivity strikes*. The difference is analogous to that between fasting for personal reasons and engaging in a hunger strike. Both involve voluntarily refusing to eat. In a hunger strike, however, this refusal is keyed to public demands for specific concessions from those in a position to grant them. The specific and public nature of the demands is crucial to strike efficacy. But here the analogy with connectivity striking breaks down. Hunger strikes can be effectively carried out by individuals as long as they are able to stir humanitarian guilt on behalf of those empowered to grant sought-after concessions. Connectivity strikes, like labor strikes, must be undertaken as mass actions to be effective, and the leverage they exert must depend on something other than the guilt of the powerful as fulcrum.

To take up a specific example, consider how an organized connectivity strike might be used to counter the kind of algorithmic biasing of electoral dynamics that occurred in the 2016 US presidential election. The use of machine learning and AI to sway public opinion and influence voting behavior is now common political practice, with the most crucial period being the last two weeks before an election when swing voters are finalizing their commitments. If a sizeable demographic were to commit to a social media strike for those two weeks, relying for election information only on traditional print and broadcast media, along perhaps with free and fast fact-checking sites, it would greatly reduce their susceptibility to manipulation. A connectivity strike would, in effect, place the voting public in data quarantine, out of the reach of adaptive artificial agents programmed to use citizen data to craft maximally effective, behavior-shaping informational feeds. It would also send a clear message to candidates that a significant part of the voting public considers the robotic manipulation of public opinion to be a political liability.

Of course, traditional print and broadcast media are also used to shape public opinion. But as Tim Wu (2016) points out in *The Attention Merchants*, their persuasive capabilities pale in comparison with those afforded by algorithmic intelligences making use of detailed information about voter preferences, priorities, and vulnerabilities. Today, over half of global internet traffic is generated by robots. Connectivity strikes can place voting publics out of robot reach. But one can imagine going a step further. The AI tools used to craft fake news items can be used instead to edit them, removing all but policy- and platform-clarifying candidate statements. This could be used to lobby for campaigns in which politicians engage in substantive discussions of issues rather than doing their best to short-circuit critical reflection through the use of inflammatory rhetoric. Candidate-generated content that failed to meet these criteria would simply be filtered into digital oblivion.

Connectivity fasting, undertaken in support for organized strikes, could be used to gain similar leverage for changing corporate conduct. Consider, for example, how social media and e-commerce strikes undertaken by key demographic groups might be used to elicit corporate commitment to data applications aligned with strike-enunciated priorities. To date, efforts to allow consumers/users to have a say in how their data is used have focused on building regulatory frameworks that ensure informed consent. The most sweeping attempt of this kind is the new General Data Protection Regulation, which was rolled out in the European Union in May 2018—a framework that forces corporations to specify how they will use consumer/user data and that also enables consumers/users to withhold consent for any uses of their data that they find unacceptable. The impacts of this new regulatory framework for securing individual data ownership rights remain to be seen. But as noble as its ambitions are, they do not go beyond constraining corporate conduct. Connectivity strikes could be used to begin systematically *redirecting* corporate data-use behavior.

If undertaken in an organized fashion by a highly valuable demographic group like college and university students—a crucial population of early digital adopters numbering over two hundred million worldwide—connectivity strikes could be a core practice within what Tim Wu calls a "human reclamation movement" based, first, on boycotting technologies that distract, diminish, and engage in "attention theft" and, secondly, on demanding technologies that help us focus and think (Wu 2016: 351). But perhaps more importantly, connectivity strikes by college and university students or other key demographic groups could also be used to leverage research and development agendas into better alignment, for example, with realizing already-specified and widely endorsed social justice goals.

As never before, withholding attention gets attention. Until now, the informational yields of attracting and holding human attention energy have been directed overwhelmingly to advancing corporate rather than human interests. Well-organized, global connectivity strikes could be used to get the proverbial ear of corporate leadership and invite them to step forward and devote specified percentages of their research investments to exploring, for instance, how machine learning and AI might help meet the social justice goals specified by the United Nations Millennium Development Goals or the 2030 Agenda for Sustainable Development.[4] A further step could be to make "brand loyalty" contingent on corporate commitments to meeting such goals and providing consumers/users with progressively healthier connectivity diets and environments.

Unlike the strategy of corporate regulation—a strategy that arguably has the result of spurring intelligent corporate "work arounds" to secure undiminished returns on investment—connectivity strikes can be used to create conditions in which corporate leaders are able to gain and secure attention share by emulating the kind of leadership sought by striking publics. Given the recent record of industry regulation, redirecting corporate creativity in what amounts to a win-win fashion is arguably preferable to restricting it.

Considerable international policy leverage can also be exerted by global publics taking advantage of the economic impacts of connectivity strikes to gain the attention of the leadership in governments—like that of the People's Republic of China, for example—which rely crucially on economic progress for continued legitimacy. While online forums have clear limitations when it comes to articulating coherent policy platforms and carrying out practically effective collective action (Castells 2015), they have great potential for organizing boycotts of platforms and products originating in countries heavily invested in the New Great Game. Unlike top-down, state-originated tariffs on goods and services which are eventually passed on to consumers and which typically result in instrumental policy tradeoffs, connectivity strikes and product boycotts could be used to reframe national and international policymaking agendas from the ground up. Nations and cities are also competitors in the global attention economy and vulnerable to purposefully effected volatility in it.

The Limitations of Connectivity Strikes

The effectiveness of connectivity strikes depends, like that of labor strikes, on being large, clearly targeted, and temporary. If the striking public is too small, if a strike lasts too long, or if a group strikes too often, corporate incentives for working around the striking population and its demands can be very tempting. At a certain point, it simply becomes more cost-effective to replace striking data producers than to meet their demands. While strikes can be organized quickly, they directly slow industry growth only temporarily. Connectivity strikes are tactical, not strategic.

If the objective is to direct data-generated revenue away from developments of AI and data applications that place human intelligence most at risk, there are more strategic approaches to affecting the practices of leading competitors in the New Great Game. The most obvious of these, of course, is to redistribute corporate revenues by taxing profits on attention and data exploitation. Doing so, however, would require global institutions for assessing and collecting corporate income taxes. These do not currently exist and would likely take decades to build. A more innovative option is to shift the collection process from the "back end" of the production chain to the "front end" by creating conditions in which individual data producers are paid up front for rights to use their data, any time that it is used for commercial *purposes* rather than when it generates corporate *profit*.

This could be accomplished, as Jaron Lanier (2014) has suggested, through a basic redesign of the technical infrastructure of the internet that would ensure direct compensation anytime one's digitally circulated data is used, for example, to train machine learning algorithms or improve search engine performance. In effect, this redesign would guarantee that everyone who makes use of digital platforms and portals receives the equivalent of an automatically generated stream of "universal basic data income." Deciding how to use or invest this income would be entirely up to individual consumers/citizens, not to corporate or state interests.

This is a bold and in many ways appealing suggestion. But even setting aside the technological challenges and likely corporate (and state) opposition to such a technical, market-based solution, I suspect that the long-term results would differ little from those being generated today under conditions of multinational corporate dominance. A change in the *technical infrastructure* of the Intelligence Revolution without a revision of its *ethical superstructure* is not likely to be enough to redirect its dynamics toward realizing more socially just and humane global futures. In fact, treating data as a private good for which each of us deserves individual compensation could produce very compelling personal incentives to produce ever more and better data. Ironically, being compensated individually for uses of our personal data might in fact accelerate movement on the current course of the Intelligence Revolution rather than slowing or significantly redirecting it. An approach more likely to succeed in slowing and reorienting the growth dynamics of smart capitalism is to develop alternatives to data governance presumptively based on conceptions of data as private property and of commercially motivated data use as an economic activity free from negative externalities.

Data as the "Carbon" of the Attention Economy: Toward Public Good Data Governance

One such alternative, briefly introduced earlier, is to treat human-generated data as "genetic"—that is, to regard data not as personal property but as extended personal presence. Conceptually, this is appealing since it builds on a relational understanding of persons and can draw on ethical and legal resources that have been developed to treat the human genome as a global public good. But while it is worthwhile advocating for treating all human data as a global public good or resource commons, the very modest returns thus far on efforts to similarly treat just the genomic subset of human biodata suggests that other legal and ethical strategies should also be explored.[5]

A second possibility is to broaden data considerations to extend beyond personal human data of the kind that might plausibly be considered part of one's "karmic genome" and to include all the data that is currently being generated and collected since all of it, directly or indirectly, "belongs" to humanity. Although machine-generated observations of other machines' conduct produce immense amounts of data, the intentions motivating the collection of these sensory/observational atoms—time- and location-stamped values for specific variables—are nevertheless human. Taking this more comprehensive datasphere approach opens prospects for applying insights from the climate change predicament to argue on behalf of using attention- and data-focused cap-and-trade and/or tax schemes to slow the growth of the Attention Economy 2.0, much as carbon cap-and-trade and tax schemes are being employed to mitigate and/or draw down the negative externalities produced by the global carbon economy.

These two options imply different conceptions of data. This might be troubling to those wedded logically to the principle of the excluded middle—the principle that something cannot be both A and not-A. The Buddhist teaching of emptiness suggests, on the contrary, that all that exists (matter) is the definition of a value-laden point of view (what matters) allows us to see different conceptions of data as ways of opening different actionable possibilities. That is arguably a very good thing, especially in the present case. But an analogy could also be made with light, which contemporary physics is comfortable regarding as either a stream of particles (photons) or as a continuous wave, depending on the experimental context.

The Datasphere as Relational Commons

Given the practical—especially legal and regulatory—challenges of data-centered approaches to slowing the Intelligence Revolution, the broader, datasphere approach that regards data as a source of economic energy arguably has practical advantages over seeing data as genetic. Schematically, the datasphere can be thought of as having three distinct dimensions: (1) data about physical systems, including everything from biological to industrial to astrophysical systems; (2) impersonal, aggregate data about people as objective "bodies in motion," not as willing, desiring, and values-embodying subjects; and (3) personal, individuated data about humans as relationally constituted, meaning-seeking, and values- and intention-expressing subjects. Of the zettabyte of data that it is estimated will be generated every two days by 2025—enough to film eighteen million years of HDTV—the greatest portion will *not* be identifiable as specifically personal data. It will be data gathered through a highly distributed, artificial "sensorium," including a steadily expanding internet of things, which is devoted largely to tracking physical state changes and object-to-object interactions across scales ranging from the microscopic to the macrocosmic.

Although a great deal of legal and ethical attention has rightly been directed toward addressing worries about the uses and abuses of personal data (including supposedly anonymized data) or toward specific surveillance techniques like facial recognition, the resulting focus on the agents and patients of data use has tended to divert attention away from big data's more extensive and predicament-generating structural/relational ramifications. Among these has been the "network effect" by means of which technology giants have established what amount to data monopolies enabling them to reap nearly all of the immense financial benefits which accrue through them.

Granted the fact that digital platform and portal giants are reaping enormous profits from technological advances originally created with taxpayer money—including the internet, Google's search algorithm, touchscreen displays, global positioning systems, and voice-activated personal assistants like Apple's Siri—and granted further the fact that their business model is based on taking advantage of the habits and private information of the taxpayers who funded these advances in the first place, the argument has been made that the public's data would be more justly managed if was "held" in a public repository for sale or lease to commercial interests. This would not only enable a portion of the profits from data to be directed back to citizens, it would allow the public to play a role in shaping the digital economy to satisfy public needs. In short, big data might best be conceived of as a digital public good (Mazzucato 2018).

In contrast with the tactical nature of connectivity strikes, reconceiving digital data as a public good opens strategic prospects for treating data aggregators and owners as responsible for serving the public interest, not simply their own. In practice, this might mean, for example, requiring companies to contribute a percentage of their annual revenue to publicly managed funds or limiting corporate rights to profit exclusively from publicly generated data by establishing a fixed term for proprietary data use.

One of the challenges of this approach, however, is that the complexity of the datasphere as a commercially valuable ecosystem of objects, people, and privately built and maintained digital platforms and connectivity infrastructures makes it difficult to map within the compass of the traditional concept of a public good. Although the most commercially valuable means of digital data gathering and analysis might originally have been developed through publicly funded research, they are now owned almost entirely privately (Purtova 2015). Any discussion about treating big data as a public good is thus unavoidably a discussion about acknowledging and resolving conflicting interests regarding both power over data ownership and power over the uses to which data can be profitably put, regardless of who owns it (Taylor 2016).

These caveats notwithstanding, there is still considerable critical promise in seeing big data as a global public good and the datasphere as a global resource commons, focusing strategically on data governance as a means to resisting the headlong acceleration of the Intelligence Revolution and setting it on a more humane and equitable course. Traditionally defined, public goods are goods, services, or institutional arrangements that are: (1) beneficial both to individuals and to society as a whole, (2) nonexcludable in terms of how their benefits are distributed, (3) nonrival in terms of their use or consumption, and (4) subject to market failure or under-provision. Education is a classic example of a public good. Commons are domains from which generally useful resources or goods cannot be drawn long-term unless collectively governed. Community forests, lakes, and ecosystems are classic examples of resource commons. The datasphere seems to straddle these categories in ways similar to geo-stable orbits and the internet, both of which have come to be conceived as both global commons (GC) and global public goods (GPG): that is, technologically realized environments that facilitate interactions and afford private and public benefits across borders, populations, and generations. We might call these global *relational* commons.

The possibility of regarding environments—whether natural or artificial—as collectively governed resource commons, the conservation of which serves long-term public goods purposes, can be traced conceptually to Garrett Hardin's seminal paper "The Tragedy of the Commons" (Hardin 1968). In it, Hardin argued that, without a "fundamental extension in morality," meeting the needs of a rapidly expanding global population would eventually come at the cost of the degradation and eventual collapse both of global resource commons and of conditions for the continued pursuit of human flourishing. In recommending a merger of quantitative (scale) and qualitative (ethical or moral) concerns to secure the integrity and sustained viability of global ecological systems and societal well-being, Hardin can be read as implicitly directing attention away from conserving seemingly independent *pools* of already existing material resources to conserving patterns of productive *processes*. That is, he can be read as blurring the conceptual boundary between resource commons and public goods in ways that compel bringing economic, environmental, and ethical values into practical alignment.

Extending the concept of the commons from a static pool of available resources to encompass dynamically productive and value-generating processes implies, for example, that the continued viability and vitality of the global "knowledge commons" depends on providing education as *a* public good in ways that are inseparable from pursuing *the* public good through education.[6] In effect, this marks a fusion of the traditional economic conception of public goods as responses to market failures with ethical commitments to correcting for the partial—that is, both incomplete and biased—nature of so-called free market solutions. In short, the provision of public goods goes beyond playing a merely *compensatory* economic role to playing an increasingly normative or *evaluative* one. This resonates well with the Buddhist insistence on the continuity of metaphysics and ethics or the "space of causes" and the "space of reasons."

It is possible, then, to see the datasphere as a global relational commons that is realized by means of a dynamic, privately-owned and -operated network infrastructure that performs a public good function of intelligence gathering. A crucial advantage of doing so is that it opens prospects for moving away from adversarial relationships among the public, governments, and digital platform/portal companies and into more positive, collaborative, and incentives-based relationships. Ideally, this move would enable these companies to "derive the highest possible value out of the data that they have amassed to address a wide variety of societal challenges, without infringing the legitimate interests of both platform companies and their users" (Shkabatur 2019: 411). The long-term goal of data governance is thus guaranteeing that corporate-generated data is made available as *a* global public good that can be capitalized on in scientific, social, cultural, and economic contributions to enhancing *the* public good.

Data Governance in the Attention Economy: The Risk of Unacknowledged Externalities

There are, however, two critical concerns about conceiving of the datasphere as a global public good and relational commons. First is its implicit validation of assuming that the bigger the data or knowledge commons, the better. Second is the way in which it tacitly directs critical regard away from the negative externalities associated with the capture and exploitation of data-conveying and intelligence-revealing human attention, suggesting that data generation is a cost-neutral process.

If the datasphere is a global commons on which all of humanity can draw as a pool of basic (epistemic) resources to direct and sustain data-driven public goods processes, it stands to reason that the larger the data commons becomes, the greater will be the benefits it affords. There are two problems with this conclusion. The first has to do with data and attention quality. The "genetic," karma-encoding data we produce purposely (e.g., through social media posts) or inadvertently (e.g., through leaving internet cookie trails) can contain evidence of *kuśala* conduct, but that is not generally the case.

Recalling the Buddhist distinction between attention captured by the superficial, craving-inducing aspects of things (*ayoniśomanasikāra*) and attention that is concentrated and directly freely toward *truing* relational patterns (*yoniśomanasikāra*), it is the former that is characteristic of the attention that is most powerfully exploited in the Attention Economy 2.0. It is also the kind/quality of attention that is associated with the "polluting outflows" (*āsrava*) generated by self-affirming views, inapt habit formations, sense attachments, and clinging desires—the kind/quality of attention that causes, perpetuates and intensifies conflict, trouble and suffering. Amassing the data conveyed by this kind/quality of attention and using it to train machine learning algorithms risks bringing about conditions conducive to runaway data pollution. And the same can be said of algorithmic agencies directed to spread specific kinds of disinformation or otherwise consciousness-degrading digital content.

Yet, even setting aside data quality concerns, a bigger data commons is not necessarily better. Even if a commons allows for universal access, the distribution of the benefits of access will not necessarily be equal. This is the second problem. A small forest may serve well as a commons since it is readily accessible in its entirety by all nearby communities; it yields its resources in ways that disadvantage none; and it affords relatively quick and clear evidence of emerging patterns of degradation or overuse. But as the size of a commons increases, collective governance becomes increasingly difficult. It can be hard to bring geographically distant user communities to face-to-face deliberations, which can in turn result in the emergence and/or exacerbation of value differences regarding the "proper" uses of the commons.

Even if these difficulties can be overcome, the "larger" a commons becomes, the more difficult it becomes to access the entirety of its resources. Reaching the interior of a very large forest where the oldest growth and most valuable trees exist might be possible only with considerable material investment—pack animals, provisions, a large labor force, etc. That is, the benefits of "equal" access naturally accrue more readily to those who possess the means and power to access the "interior" of the commons. It is already clear, for example, that so-called universal access to such technologically realized relational commons as geo-stable orbits or the internet has *not* ensured that the benefits of access are evenly distributed. It is big investors who win big. An exponentially growing data commons is similarly prone to disproportionately advantage those who are able to draw on its resources both extensively and deeply.

The argument might be made that this really should not matter. The constant production of new data means that big data is an effectively "inexhaustible" resource pool and a public good that will always be in ample supply. Guaranteeing the sustainability of big data use is not an issue. The data environment is not in danger of exhaustion or degradation. Big data thus constitutes a good that is almost ideally nonexclusive in terms of its use and that supports theoretically unlimited, nonrival benefits. Yet, just as people with home libraries, computers, and tutors can make better use of a public education than those who lack them, whoever has the most extensive data-analysis capabilities is able to make the most extensive and impactful uses of big data. Ironically, the equity challenges associated with treating data as private goods can reappear and intensify in public goods approaches to governing the global data commons. In the end, we are brought back to evaluative considerations of the purposes for which data is utilized.

It might well be that at some point in the future, data will be collectively governed through global institutions that guarantee the fair distribution of benefits. Perhaps the principle now being promoted to govern the access and use of genomic data will become adopted as institutional norms for overall data governance. Unfortunately, these institutions and principles remain aspirational. Given the track record of developing globally honored commitments to lowering carbon emissions and resolving the climate change predicament, and given the fact that competitive advantage in the New Great Game is so tightly linked to being at the cutting edge of technological innovation, it is hard to imagine that comprehensive and effective data-governance institutions will be built in time. If the Intelligence Revolution is to be slowed, more immediately impactful governance is needed.

One possibility is to revisit and take critical advantage of the role of attention as the "conveyance" of human data flows into the intelligence-gathering infrastructure of smart capitalism—data flows without which algorithmic innovation is significantly hobbled. Instead of focusing on short-term uses of data-derived leverage in connectivity strikes, however, longer-term critical advantage might be gained by pressing the metaphor of attention—funneled through the computational factories of the Fourth Industrial Revolution—as the energy or fuel powering the global production and circulation of goods and services. Going a step further, drawing an analogy with the role of fossil fuels in powering the machines of earlier industrial revolutions, prospects open for repurposing the carbon cap-and-trade and carbon tax strategies that have been developed to slow anthropogenic climate change and guard against pollution damage to natural ecosystems—that is, prospects open for developing attention/data cap-and-trade and attention/data tax regimes to slow the Fourth Industrial Revolution and mitigate its impacts on the anthrosphere.

Rather than attempting to do the ethical and diplomatic work needed to resolve the value conflicts made manifest by anthropogenic climate disruption, carbon cap-and-trade and carbon tax regimes are practically focused attempts to slow industrial uses of carbon-based fuels and lower greenhouse gas emissions. Carbon cap-and-trade schemes regulate carbon dioxide and other emissions by setting industry-wide caps on allowable emissions and distributing emission allowances that can be used or traded by companies making use of carbon-based fuels. This gives companies the option either to work within their allowance or to buy a share of other companies' allowances to be able to maintain their current energy practices.

Carbon tax schemes lower emissions by creating clear cost incentives to cut carbon fuel use and/or develop energy alternatives. The logic here is not only to limit emissions but to quantify the social and environmental costs of carbon fuel use.

Limiting carbon emissions is the most direct means of lessening the scale and impacts of the negative externalities associated with climate change. One of the difficulties of realizing comparable cap-and-trade or tax schemes for attention/data is the current lack of consensus on precisely what kinds of polluting "emissions" and negative externalities result from big data generation and use. Claims about job loss due to data-driven smart services, robotics, and AI can be countered by claims about their job creation potentials. Claims about the disruptive impacts of social media on existing conceptions of family, friendship, and citizenship can be countered by claims about how social media expand freedoms of expression and identity formation and facilitate new and more flexible forms of collaboration. Attention/data caps and taxes are powerless strategically in the absence of robust global consensus on the problematic "emissions" to be curbed by them.

Capping or taxing the carbon emissions involved in energy production and use makes sense. Carbon-based fuels do not become an economically relevant source of energy until they are exploited—that is, burned. Mining coal, oil, and natural gas does have direct environmental costs, but these are dwarfed by those caused by converting their chemical potential energy into mechanical energy. In the Attention Economy 2.0, however, negative externalities do not result first or only with attention/data exploitation; they are present already in the initial extraction of attention/data taking place in our homes, schools, places of work, and sites of leisure—anywhere touchscreens, smartspeakers, and voice-activated wireless devices are present. When attention is captured and subjected to computational analysis to extract the data it conveys, it is less like mining and burning fossil fuels than it is like cutting down and burning forests. The environmental impacts are immediate. Cutting down a forest is not simply an act of felling a multitude of individual trees, it is cutting through—and potentially leaving in tatters—the relational fabric of a forest ecosystem. The same is true of harvesting attention. The new attention industries are not merely extracting and exploiting your time and data or my time and data, they are cutting through the relational fabric of attention present both in the social processes by means of which value is produced through in-the-moment imitations and improvisations, and in the shared values, intentions, desires, beliefs and emotions in which they result.[7] The disruptive impacts on social ecologies—and on the distinctive patterns of social coherence and resilience that characterize them—are immediate. If all things are what they mean to and for one another, the karmic effects of harvesting attention are necessarily as social as they are personal.

It should be stressed that this reading of the effects of attention extraction and exploitation on human sociality is not inconsistent with digital connectivity facilitating the emergence of new and perhaps quite positive social arrangements and movements. The very possibility of organizing connectivity strikes depends on that potential. Yet, as is the case with the Intelligence Revolution as a whole, the dynamics of the Attention Economy 2.0 are predicament laden. Cutting down forests helped fuel early phases of the modernization and industrialization processes that have contributed to much-improved life quality for much of humanity, as well as the emergence of new forms of human agency in a marvelously expanding range of socially constructed (virtual) realities. That has arguably been very good for humanity. Yet, there is no simple metric for determining whether humanity's gains have proven to be "worth" the ecological damage thus incurred. The progress that has been made may be good, but it has not been demonstrably superlative.

The same has been true thus far of the harvesting of human attention. As the karmic critique of smart services and their "unintended" consequence of attenuating intelligent human practices has hopefully made clear, on its current heading the Intelligence Revolution will place us at risk of forfeiting precisely the skills needed to successfully engage in the predicament-resolving ethical labor of human course correction: our arrival at an ethical singularity. That seems to me an eventuality eminently worthy of avoiding. There are times when good outcomes and opportunities are not enough.

But there clearly are people who continue to see the natural world as having no value other than human utility and who regard ecological damage as something to be concerned about only if it can be proven to have serious disutility for humans. Comparably, there are transhumanists and those convinced of the (at least potentially) superior rationality of algorithmically tuned markets and governments who regard concerns about the "forest clearing" needed to arrive at truly smart futures as evidence of nothing more than human chauvinism. While a globally enforced data cap-and-trade regime could serve as an effective means of limiting overall attention extraction and exploitation, it would face the same challenges of establishing effective global values consensus as carbon cap-and-trade strategies. These challenges are not insurmountable. But it is almost certain that efforts to meet them would be countered by Big Tech and the current beneficiaries of what amount to data monopolies. Taking a lesson from the playbook for reducing carbon emissions, something like a blended or hybrid approach combining features of cap-and-trade and tax approaches is likely to be most effective.

A practical advantage of an attention/data tax strategy for slowing the Intelligence Revolution is that it does not require a Big Tech breakup. The political and institutional challenges involved would not be incidental. Global deliberations involving as wide a range of stakeholders as possible would have to be undertaken to determine both tax rates and how best to use resulting tax revenues, organized perhaps by an intergovernmental organization like the United Nations. Institutional mechanisms would then have to be put in place to monitor how much attention/data is being acquired and used by digital platforms and portals, the purveyors of smart services, and algorithm-assisted e-businesses, as well as mechanisms for enforcing payment. The revenues thus generated would need to be distributed equitably and with some consensus regarding how to use them to mitigate or protect against the damage to social ecologies caused by attention/data extraction.

These challenges notwithstanding, the advantage of taxing attention/data *extraction* is that it circumvents the need to determine in advance what the most powerfully negative externalities of attention/data *exploitation* will be in the long run. This has been one of the stumbling blocks of the cap-and-trade approach to curbing carbon emissions: the need to get very different groups of stakeholders to agree on which of the risks and likely impacts of continuing "business as usual" are too high to ignore. An attention/data tax, if sufficiently high and progressive, would also fairly incentivize commercial caution—especially among small- and medium-sized businesses—in taking advantage of digital business services or "retailing" smart systems since platform and portal companies will almost certainly pass on as much of their own tax burdens as possible to those further down the data exploitation chain.

In light of the fact that attention/data extraction is *not* a coercive practice, however, but rather a voluntary arrangement by means of which consumers/users exchange data for access to desirable goods, services, and experiences, there is a case to be made that consumers/users should also be subject to an attention/data tax. The logic here is similar to the excise taxes on alcohol and tobacco that are currently levied by governments as a way of both slowing consumption rates and offsetting associated healthcare costs and productivity losses. A "digital excise tax" would have the effect of moderating consumer/user expenditures of attention and time and could be used to fund programs aimed at fostering personal creativity and community resilience.

There are, of course, a number of arguments against excise taxes, including arguments that these so-called sin taxes increase unemployment, that they are regressive in having greater impacts on the poor, and that they incentivize tax avoidance and evasion. Strong global evidence exists, however, that these arguments are either invalid or compromised by greatly exaggerated claims for the generality of their factual premises (Chaloupka, Powell, and Warner

2019). And it is important to note that the logic of a digital excise tax does not rest on the kind of moral judgment that is associated with taxing alcoholic beverages but not, for example, sugar-heavy "fruit-flavored" drinks. On the contrary, to be most effective, a digital excise tax would have to be content neutral. It is not a means to directing attention away, for instance, from pornography or gambling sites. Its purpose is not to discourage or encourage certain types of digital connection but rather to offset the "opportunity costs" of digitally accomplished attention capture.[8]

Engaging in connectivity strikes, efforts to institutionalize global commitments to regarding the datasphere as a global public good and relational commons, and pursuing the imposition of attention/data cap-and-trade and/or tax regimes are complementary approaches to resisting and slowing the change dynamics of the Intelligence Revolution. Each of them appeals to different conceptions of attention and data, bringing into focus distinct sets of actionable possibilities. In Buddhist terms, they are different skillful means (*upāya*) for changing the way things are changing. Implicit to each are profoundly ethical concerns about how attention and data should be invested—concerns about redirecting the dynamics of the human-intelligent-technology-world relationship—and their success ultimately depends on being partnered with parallel efforts to develop the ethical intelligence and diversity needed to address those concerns and to share in resolving the predicament of intelligent technology. To bear practical fruit, these ethical efforts depend, in turn, on being informed by high-resolution understandings of the historical currents shaping the present moment, of the workings of intelligent technology, and of the possibilities for engineering course change aspirations into concretely embodied collective action. Redirecting the Intelligence Revolution will depend on developing new ecologies of knowledge in improvisation-fostering and predicament-resolving educational environments.

Course Correction: A Middle Path Through Education

Two popular convictions about education are often forwarded today by politicians and supported by some educational administrators and future-oriented education theorists. The first is that educational reform is desperately needed to meet real-world demands. The second is that science, technology, engineering, and mathematics—that is, so-called STEM disciplines—should be at the heart of the educational experience. Otherwise, the argument goes, students will continue to graduate with "worthless" degrees in humanities disciplines like history, literature, and philosophy, shackled with debt, and with earning prospects well below those of their peers who concentrated their studies in STEM disciplines.

The plausibility of the second conviction depends fundamentally on a failure to fully appreciate the actual needs of today's employers (many of whom greatly value the kinds of communication and critical thinking skills that are the hallmark of a liberal arts education) and a failure to adequately anticipate the labor market impacts of intelligent technology. Not only will tomorrow's computational factories have no need for human machine operators or maintenance mechanics, they will also have no need for mechanical, chemical, and electrical engineers. As the research, reasoning, and innovation capabilities of artificial intelligence systems develop and mature, algorithmic agencies will assume responsibility for increasing proportions of the intellectual labor involved in science, finance, medicine, and law.

But the plausibility of calls for STEM-only education also depends on a failure to appreciate the revolutionary character of the already well-advanced transition from a human-technology-world relationship premised on the primacy of problem-solution to one shaped by the growing necessity of predicament-resolution: a transition from a relationship in which the illusion of independence and the adequacy of linear causal explanations was sustainable to one in which the irreducibility of interdependence and the prevalence of nonlinear causalities is deniable only on the pain of growing practical incompetence. Not only are we not in need of education framed around a radically trimmed back curriculum, we are on the verge of no longer needing a standards-based, curriculum-organized approach to teaching and learning.

Improvisational Readiness: The Essential Imperative of Twenty-First-Century Education

Indeed there has been considerable pushback against the conviction that the twenty-first century education should be STEM-centric. Even if the negative impacts of the Intelligence Revolution on labor markets are not as severe as currently predicted, replacing liberal arts study with STEM-only education would still be misaligned with both personal and societal needs. Well-argued and data-supported cases are being made for more effectively integrating the study of liberal arts and STEM disciplines to foster real-world innovation and communications skills (Anders 2017), to ensure that democratic politics and the pursuit of global social justice remain viable practices (Nussbaum 2016), and to promote the kind of creativity and passion for discovery that will be needed to prosper personally and contribute societally in a world transformed by artificial intelligence (Aoun 2017).

STEM-dominant education is well-suited to producing graduates with strong problem-solving skills and a penchant for innovation in narrowly bounded knowledge applications. It is not as well-suited to fostering the kinds of predicament-resolving facility, open creativity, and improvisational readiness that are needed to address the ethical—not merely technical—challenges that are being posed by the Intelligence Revolution. To meet those challenges and to resolve global predicaments like climate change, training in the humanities and social sciences is essential. But simply ensuring that this training continues to be a crucial part of education and that it is carried out in ways which are student-centered and orientated toward active learning will not be enough. In addition, humanities and social science teaching and learning—along with teaching and learning in STEM fields—will need to be substantially reoriented as intelligent human practices. We are in need of a new educational karma: a new constellation of educational values, intentions, and practices.

Curricular Education: An Artifact of Industrial Modernity

Some educational history is useful in explaining this need. The dominant paradigm of formal education is a distinctively modern institution. Its origins can be dated back to 1576 when the French philosopher and mathematician

Peter Ramus used the term "curriculum" to epitomize a radical alternative to the model of education that was dominant at the time: a studio-centered, apprenticeship model of learning-through-participation. Against this model and its presumption that learning occurred at individually appropriate rates over indeterminate periods of time, Ramus claimed, first, that knowledge is a quantifiable and deliverable good which is most efficiently and effectively transferred via standard instructional sequences and second, that education consists in transferring specific bodies of knowledge and competence. Whereas the apprenticeship model of potentially lifelong learning aimed at mastering and perhaps eventually setting new standards of practice in an intergenerational community of practice, curricular education was conceived as an explicitly terminal endeavor—a conception neatly encapsulated in Ramus' novel use of the Latin word *curriculum*, which originally referred to a circular course of standard length for chariot racing competitions. According to the curricular model, education not only can but *should* be understood as a finite game played to win.

With a nod toward Ramus' renown as a mathematician, his curricular alternative to studio-based teaching and learning could also be called education by algorithm. This standardized and course-based approach to education proved highly compatible with the processes of nation-state building and industrialization—so much so that Adam Smith, the father of market economics, insisted that mass education should be provided by the state as a core "public good" in the interest of developing the kinds of citizens needed to stimulate and sustain nation- and empire-building. Wherever mass education of this kind has been implemented, the results have been unequivocal: greater productivity, increasing per capita incomes, rationalized population growth, labor market differentiation, and the inculcation of "civic" sentiment. Indeed the compatibility of curricular education and nineteenth and twentieth century processes of industrialization was so profound that it seemed entirely appropriate to regard schools as "factories," students as "raw material," and graduates as ready-to-use "finished products."[1]

Today, the blunt instrumentality invoked by these metaphors and the apparent denial of learners' agency will strike most of us as socially misguided, if not morally offensive. Tellingly, it is also a set of metaphors that seems ill-matched to contemporary economic realities and incapable of offering useful insights for educational practice and reform. Ours is a world in which the economic logic of accumulation has given way to a logic of circulation in which the profits of centralized industrial factory production are shrinking rapidly in relation to those generated by decentralized networks of innovation and design studios and in which the values of manufactured products and real property are being progressively dwarfed by those of goods and services prototypes and intellectual property (Lash, 2002). As successful as the modern paradigm of algorithmically methodized curricular education proved to be, it is not a viable way forward.

Discipline-focused, grade-incentivized, standards-driven, and course-delivered education not only fails to deliver workforce-ready learners, especially in times of rapid economic and labor market change, it fails to help them become truly world ready. A broad consensus exists among educators that adapting education to address the disruptions being brought about by the Intelligence Revolution will require an educational revolution: a "turning back" toward key features of the premodern, studio apprenticeship model of flexible, individualized, and continuous or lifelong learning, but in full cognizance (and taking full advantage) of the technological transformations that have occurred in the intervening five centuries.

Educating for Anticipated Futures

The strategies for embarking on this educational revolution tend to converge on reconfiguring institutional structures to enable the adaptive embedding of educational practices in the rapidly—and often unpredictably—changing dynamics of contemporary societies. That is they are strategies expressing some degree of confidence in our forecasting abilities. Two of the more notable approaches for doing so in higher education are those forwarded by Cathy Davidson (2017) in *The New Education: How to Revolutionize the University to Prepare Students for a World in Flux* and by Joseph Aoun (2017) in *Robot-Proof: Higher Education in the Age of Artificial Intelligence*. Among their recommendations for reform are restructuring colleges and universities to promote issue-responsive rather than generic teaching and learning; to foster broad bandwidth experiential learning through extending education into relevant communities, globally as well as locally; to reward the kind of divergent thinking needed to produce and evaluate multiple answers rather than the convergent thinking promoted by "right answer" testing; to engender facility in systems and critical thinking; and to enhance entrepreneurship and cultural agility.

These are valuable recommendations. Implementing them would certainly help bring about a shift in educational emphasis from transferring existing bodies of knowledge to enhancing readiness for engaging in adaptive conduct—a reorientation of teaching and learning from transmitting and acquiring knowledge to practicing (and not merely exercising) intelligence. But to be effective in helping cultivate the improvisational virtuosity that is needed practically to resist the colonization of consciousness and ethically to resolve the intelligence predicament, the purposes of this reorientation will have to be further qualified.

In *Not For Profit: Why Democracy Needs the Humanities*, Martha Nussbaum (2016) argues passionately and persuasively on behalf of a shift of educational energy away from securing national economic gain to ensuring that educational processes promote the kinds of personal moral development, compassionate concern, and imagination needed to pursue social justice and to contribute to securing the conditions of democratic governance. The United Nations Educational, Scientific and Cultural Organization (UNESCO) forwards a compatible vision of global citizen education in its *Education 2030 Agenda and Framework for Action*, calling for all countries to ensure that every learner develops skills for reflecting on and promoting sustainable development, human rights, gender equality, peace, and nonviolence, while at the same time ensuring that learners understand their citizen rights and responsibilities in the context of their own countries and cultures. In both cases, a vision of education is being forwarded that ultimately prepares young people to deal effectively and empathetically with the complexities of living together interdependently.

The stress that Nussbaum and the UNESCO place on citizenship-relevant skills for addressing the challenges of living together in complex societies has considerable merit as a qualification of the calls made by Davidson and Aoun for educational reforms aimed at engendering workplace-relevant creativity. At least in principle, it is a stress that intimates the importance of ethical intelligence and improvisation. But, given the populist turn in American, European, South American, and South Asian democracies, and the limited embrace of democratic values in countries like China and Russia—countries that are and will continue to be major players in the New Great Game—it is not at all clear that this qualification will suffice to do the work needed to realize practically robust educational commitments to global predicament resolution. Greater values precision seems necessary.

This need becomes dramatically more evident when the prospects of deepening incursions of artificial intelligence and smart services into the education sector itself are taken into account. As was discussed earlier, there is considerable movement toward using machine learning and artificial intelligence to provide smart educational services that will streamline administration, provide flexibly responsive instruction and tutoring, assist in grading tests and papers, and offer personalized, data-driven feedback on teaching and learning. As in other domains of intelligent human practice, the concern is that as smart educational services become more sophisticated, the slippery slope from supplementing human capabilities to supplanting them will become irresistibly steep, especially in the face of ever-mounting pressures to lower education costs. This will be especially true in the post-pandemic world.

Intelligent technology and connectivity giants have aggressively positioned themselves to capitalize on the opportunity spaces opened by pandemic response measures to establish new public–private partnerships, including partnerships for educational reform (Klein 2020). In fact, the argument can be made that so-called Big Tech is compelled to get into education for the same reasons that it is compelled to get into military and defense work: the revenue bases of global tech leaders like Apple, Microsoft, Google, Facebook, Amazon, Tencent, Alibaba, and Huawei are so large that only a handful of sectors can support their further growth at rates attractive enough to retain and gain investors.[2]

Granted that schools are primary environments for socialization, for strengthening skills in social and empathetic learning, and for fostering social cohesion and resilience, deepening incursions of commercial interests and market forces into education comes with considerable risk. The public good purposes of education are interpreted and pursued differently. Often this amounts enacting and ensuring the continued viability of a specific political or ideological agenda. These differences can be seen as troubling or as openings for comparative insight. More unequivocally troubling is the singular vigor with which corporations worldwide—tempered only in limited degree by commitments to social responsibility—enact commitments first and foremost to maximizing profit and shareholder value. This has almost invariably meant both seeking commanding market share and engineering greater market differentiation. Market-competitive corporate conduct is not structured to serve the public good except in the most abstract sense in which all economic activity can be considered beneficial to the public.

Thus, while the Intelligence Revolution is almost certain to beget a "fourth education revolution," it is uncertain whether the outcome will be the provision of quality education to all or the further stratification of educational inequalities, or whether it will result in the liberation or the "infantilization" of humanity (Seldon and Abidoye 2018). The most important factors in setting the course of educational change will be the values used to frame the purposes of education and the degree to which these values are embodied and extended in its provision.

Educating for Diversity, Equity, Humane Creativity, and Freedom-of-Attention

In discussing the six *pāramītas* as open-ended ideals of personal virtuosity, the notion of values as "modalities of relational appreciation" was introduced as a way of emphasizing the inseparability of metaphysics and ethics in Buddhism—the ultimate nonduality of facts and values or the space of causes and the space of reasons. This conception of values is crucial to framing educational resistance to and redirection of the current dynamics of the human-intelligent-technology-world relationship. Values are not abstract. They consist in actual and ongoing concentrations and extensions of appreciatively qualified patterns of relationality.

In the face of the progressive colonization of consciousness, the individualized algorithmic tailoring of the connective experience, the monopoly-biased network structure in the Attention Economy 2.0 and the human deskilling potential of smart services, there is much to recommend diversity and equity as cardinal values for reorganizing the provision of education. This is especially true given the needs for international, intercultural, and intergenerational deliberation in response to the challenges of the Intelligence Revolution and the looming transformation of education into a smart service growth market. Educating for humane creativity and freedom-of-attention speaks to the need for rethinking the purposes of education, resisting its reduction to methodized knowledge delivery and redirecting teaching and learning toward standards-advancing knowledge discovery.

Educating for Diversity. Educating for diversity begins with refusing to see diversity in purely quantitative and legislatively enforceable terms, as is now common in educational institutions. It continues with refraining from identifying diversity initiatives as having the goal of tolerance for differences in culture, religion, ethnicity, or gender. Tolerance is undoubtedly better than intolerance, but it is not enough. As suggested earlier, by conceiving of diversity as a relational quality in contrast with mere variety, valuing diversity can be seen as consisting in the open-ended practice of engaging differences as spaces of opportunity for mutual contribution.

At an institutional level, a necessary and concrete step toward doing so is to foster *interdisciplinary* teaching and learning as processes aimed at transforming the academic world from within, shifting emphasis from transmitting predetermined "*bodies* of knowledge" to eliciting creative collaborations of the kind conducive to the emergence of critically dynamic "*ecologies* of knowledge." This means conserving differences among disciplinary aims and methodologies, but doing so in ways that are progressively sensitive to potentials for significant and novel contributions across disciplinary boundaries. This is especially important in addressing predicaments like climate disruption, the persistence of hunger, and the interfusion of promises and perils animating the Intelligence Revolution—predicaments that cannot be resolved without working across disciplinary (as well as national and other positional) boundaries to gain shared clarity about the conflicts of values, intentions and actions that they make evident.

It is now empirically evident that cognitively diverse groups are able to outperform groups of the best individuals at resolving predicaments and at solving paradigm-challenging problems, and that they do so in part because diverse groups more easily avoid presupposition-mortared impasses than groups of smart individuals who tended to think similarly (Page 2007). That is diverse groups enjoy greater degrees of investigative and responsive freedom than more homogenous groups. Interdisciplinary education induces and reinforces both intellectual and methodological flexibility.

Educating for diversity also means reenvisioning *internationalization*. Especially in US colleges and universities, internationalization is often seen primarily as a means to revenue generation and to offering students a (presumably desirable and tuition-warranting) cosmopolitan campus experience. These are perfectly good goals. To the extent that internationalization is pursued as a way of fostering conditions for the emergence of diversity as a distinctive relational quality, it can serve more importantly as a means of transforming educational institutions into cultural and cognitive ecotones. College and university campuses are natural laboratories for predicament-resolving practice. If internationalization is undertaken in ways conducive to the emergence of cultural and cognitive diversity, a college or university campus can function as an organic nexus of perspectives that affords multi-scale and high-resolution insight into global predicaments and other issues of public concern, while also nurturing the breadth and depth of shared resolve that are needed to respond effectively to complex global challenges.

Importantly, this repurposing of the campus can occur across a full spectrum of the relational dynamics informing higher education, ranging from the dynamics of dormitory life to those that are generated in the classroom and in extra-curricular forums when people critically engage "the same" phenomenon from dramatically different experiential perspectives. Students from Pacific atolls or from Bangladesh (where over a hundred million people live within a few meters of sea level) will have very different perspectives on sea level rise than students from the American Midwest, just as American university students who have done military service in Iraq and students of refugee parents from the region will bring distinct perspectives to evaluating the historical realities and future prospects of geopolitics in the Middle East. Valuing these differences is part of a diversity-oriented and truly international education.

Educating for diversity will also entail making efforts to ensure that teaching and learning are significantly *intercultural*. While international student bodies and faculties are likely to be culturally varied, that does not necessarily result in them being intercultural to any educationally significant degree. The educational value of multiculturalism is a function of real confrontations with cultural difference. People who were raised on different continents but in families belonging to the transnational capitalist class, for example, can have much more in common culturally than they do with less privileged citizens in their own home countries. To become educationally significant, engaging those from other cultures cannot be casual or entirely comfortable.

The Japanese cultural theorist Naoki Sakai (1997) has argued that cultural difference is not an observable, objective phenomenon but rather an interpersonally experienced feeling of dismay, discomfort. and disjunction. Experiencing and working through cultural differences—as has been done in the "cultural ecotones" that have emerged historically in association with regional and global trading centers around the world—is strongly correlated with enhanced cultural vitality and arguably with the rise of truly ethical (rather than group-specific moral) thinking. It is in such spaces of ongoing, long-term cultural interaction that it becomes necessary to engage in consensus-generating evaluations of competing customs, uncommon linguistic practices, contrary bodies of commonsense, conflicting religious convictions, and competing rationalities.

More importantly, perhaps, in considering the value of intercultural dynamics in education, there is now considerable evidence that living and studying in multicultural environments for extended periods of time significantly enhances capacities for recruiting foreign ideas, retrieving unconventional knowledge, and both generating and exploring new ideas (Leung et. al. 2008). Sustained multicultural experiences foster the kind of creativity needed to go beyond simply differing from others to differing for them in the ways needed for the emergence of true diversity and for equitable, paradigm-challenging predicament resolution.

One of the shortcomings of the curriculum model of education is grade stratification and the presumption that education is a competitive, finite game of sorting out teaching and learning winners and losers. The speed of the Intelligence Revolution does not afford us the luxury of taking a generation or two to resolve the predicament at its heart. But generational differences are hugely important in addressing ethical concerns about phenomena—like digital connectivity and socialization—that are not uniformly experienced and evaluated by those within, much less across, generations. Educating for diversity also involves bringing *intergenerational* engagement robustly into the educational experience, whether through service learning, mentoring, or other programs that purposely cut across generational boundaries. This is especially important in technical fields where significant "generation gaps" can be as small as four or five years.

In sum, educating for diversity is ideally carried out along four dimensions of inclusion: disciplinary, national, cultural, and generational. The prospects of realizing the diversity dividend depends relationally on the qualities of inclusion attained in each. In educating for diversity, equity matters.

Educating for Equity. Educating for equity begins with resisting the reduction of equity to access or reaching the goal of comprehensive nonexclusion. Although "education for all" is a laudable goal, it is not a substitute for "educational quality for all." Equity, like diversity, can be conceived in irreducibly qualitative and open-ended relational terms. Appeals to equity now generally blend appeals to universality (common principles and ideals) and particularity (uncommon identities, practices and realities), based on the underlying presumption that the individual (human being, ethnic group, gender group, nation, etc.) is the natural and proper unit of analysis (e.g., ethically, legally, economically, or politically). Pursuing equity thus involves comparing individual life prospects and compensating for inequalities resulting from structural/institutional exclusion rather than from shortfalls of individual effort. That is, equity is identified with realizing conditions for equal access or rights to participation, for example, in education, business, and politics. Realizing equity in this sense means achieving the fairness condition of so-called equality of opportunity.

This understanding of equity has been extraordinarily powerful in coordinating efforts to redress injustices based on differences in gender, ethnicity, and religion. Yet, even in the United States, where equality of opportunity has been vigorously promoted and sought as a core dimension of social justice for over two hundred years, equal work still does not result in equal pay for men and women, and minorities continue to be underrepresented in corporate boardrooms, in high-pay and high-prestige professions, and in politics. Concepts have lifetimes. And it may be that the individual-focused concept of equity as comparative equality of opportunity is not suited to doing the work of relational revision that will be needed to address the inequalities that are being generated by smart capitalism and the colonization of consciousness.

An alternative—reflecting the inclusive Buddhist ethos of compassionately alleviating or eliminating conditions that give rise to experienced conflict, trouble, and suffering—is to conceive of equity as a relational function of capacities for and commitments to practices of inclusion that are consistent with furthering personal interests in ways that are deemed valuable by others. Thus conceived, equity is not a static index of the *difference-leveling* effects of minimizing exclusion. It is a dynamic function of the *difference-appreciating* effects of enhancing *qualities* of inclusion: a function of how well social, economic, and political relations are fostering meaningful, mutual contribution. In short, pursuing greater equity entails enhancing diversity. This shift of emphasis from comparing the status of individual persons, communities, or classes to evaluating the relationally enacted dynamics of inclusionary responsiveness has important implications for education.

Pursuing relational equity involves refraining from generic engagements with differences—especially positional differences—that focus on compensating "disadvantaged" others and focusing instead on both recognizing shared responsibility for the presence of relational disadvantages and engaging in practices that engender responsive virtuosity. Educating for equity thus involves, at the "method" level, not only a shift of educational focus from learning about each other to learning from and learning for one another but also a parallel shift at the "content" level from granting centrality to knowing that (facts) to achieving situationally responsive fusions of knowing how (skills) and knowing whether (critical discernment) to pursue specific courses of action or adopt certain constellations of values. In short, educating for equity entails reimagining education as the progressive and ever more encompassing merger of knowledge with wisdom.

Institutionally, educating for and with equity can be pursued at four distinct registers: *access equity*, which concerns who can enroll in higher education, who actually does enroll, and how long they stay enrolled; *operational equity*, which concerns inclusiveness, mobility, and fairness within a given higher education institution, as well as relational quality among students, staff, faculty, and administration; *structural equity*, which concerns resource distribution within a given higher education system, compliance with equity-affecting policies, and attention to formal and informal limits on the mobility of, for instance, students and course credits within the system; and, finally, *contributory equity*, which concerns how the benefits of higher education are distributed and the degree to which higher education furthers social justice pursuits and ideals beyond its own institutional boundaries.[3]

Crucially, this relational (as opposed to comparative) conception of equity and its stress on quality of inclusion leads to seeing the pursuit of equity as open-ended. Whereas access or nonexclusion can be legally guaranteed, once and for all, quality of inclusion is a sliding scale, an "asymptotic" ideal like that of musical virtuosity. It is a direction as opposed to a destination. Stated in Buddhist terms, the pursuit of equity consists in the ongoing realization of *kuśala* or superlative arcs of change in the quality with which people and cultural traditions—but also cognitive styles and aesthetic ideals—are included within communities of shared practice. The pursuit of equity is an ever-evolving process of realizing responsive, relational virtuosity.

Humane Creativity. Valuing diversity and equity requires a basic and crucial shift from the model of curricular completion to open-ended learning. But the openness here is not merely temporal—a matter of simply continuing education over the course of a lifetime, acquiring new repertoires of knowledge along the way. I would argue that educating for diversity and equity is ultimately a qualitatively open-ended process of caring about and caring for others—a compassionate and creative practice of responsibly and responsively fostering more humane applications of knowledge, crafting more humane institutions, and realizing more humane patterns of interdependence, both within the academy and within society as a whole. This, of course, raises questions about the nature of humane and creative presence and their joint cultivation.

It has been a hallmark of late twentieth and early twenty-first century educational reforms to call for greater institutional and curricular commitments to innovation, especially in higher education. In most cases, the main purpose of doing so is to enable graduates, their future employers, and society at large to reap the economic dividends of competition-honed, entrepreneurial problem-solving. Often, this is extended to include the creativity involved in design, marketing, and advertising, as well as the creativity involved in scientific discovery and in expanding technological applications and expertise. Less often, commitments to innovation are explained in terms of the intrinsic value of the creativity involved in producing works of the visual and performing arts, literature, and music.

These are all valid rationales for promoting innovation. But expanding innovative capacities for the purpose of commercial success, for the purpose of extending scientific and technological horizons, or even for the purposes of artistic self-discovery, self-expression, and social commentary are at best tangential to the open-ended aim of embodying, embedding, enacting, and extending the improvisational capacities and commitments involved in realizing more diversity- and equity-enriching relational environments. In fact, as it is generally directed at present, educational commitment to innovation is likely to continue feeding into and accelerating the Intelligence Revolution on its current predicament-intensifying course.

A fairly standard working definition of creativity is: 1] that creativity consists in producing ideas or artifacts that are new, surprising, and either useful or valuable; and 2] that this is accomplished either by combining or blending already existing ideas and artifacts or by exploring and perhaps transforming existing conceptual spaces (Boden 2004). This account of creativity tacitly assumes that creativity is evidenced by some unique "product" that has resulted from the work of a creative "agent." It is an account that maps well onto what was referred to earlier as closed creativity.

If all things arise interdependently, however, creativity is better conceived qualitatively and relationally—as an emergent, dynamically appreciative (value-acknowledging and value-enhancing) function of responsively virtuosic interdependence. Ultimately, there are no product-resulting creative *acts*; there is only value-enhancing creative *interaction*. This is open creativity.

Seeing creativity as relationally manifest frees creativity from the bifurcating duality of the creator and the created—the exclusiveness of the agent and his/her product. And it usefully highlights the corrective implied in the statement attributed to the famous inventor and businessman Thomas Edison, that "creativity is one percent inspiration and ninety-nine percent perspiration." Creativity is *practiced*. The only way to develop capacities for improvising musically is to sustain the intelligently disciplined effort of playing music. Similarly, the readiness for engaging in shared, predicament-resolving improvisation will arise and develop only in practicing shared predicament resolution.

Predicament resolution is never easy, but it is not an arcane process. Children in free play often confront and resolve predicaments. In playing pickup team sports, for example, whoever happens to show up at the playground gets placed or chosen onto one of two teams. If the resulting teams are unevenly matched to the point that playing stops being fun for all involved, play is halted and players are reorganized to create a better balance. The predicament being resolved in this way is a simple one: while everyone values being on a winning team, everyone also values the outcome uncertainty that keeps play interesting. But resolving it is good practice for other more complex predicament-resolving efforts. In the course of play, children will also inevitably need to settle arguments about whether game rules were broken, whether "sportsmanship" norms are being violated, and even whether the rules should be changed to facilitate play. They will, in other words, be impelled to improvise. Formally organized, parent-supervised team sports protectively take affective, predicament resolution responsibilities out of the hands of the children playing. This helps ensure conflict-free play, but it also constrains the scope of ethically charged improvisation. Play becomes finite.

Search and recommendation engines go much further, taking actively imaginative curiosity out of the learning cycle, substituting individualized choice and connection outcomes for interpersonal opportunities for shared inquiry and improvisation. Search and recommendation engines function as "teleportation" devices that have the same impact on our capacities for intelligent inquiry as cars do on our physical condition. Used often enough, they bring about the conditions of intellectual "obesity" and imaginative "shortness of breath" or lack of inspiration. It is no less crucial to secure basic intellectual conditioning and curiosity at an early age than it is to ensure motor skill development and cardiovascular health.

To take the analogy a step further, if it is true that eating and exercise habits developed in childhood can be extraordinarily difficult to change later in life, it follows that educating for improvisational readiness is not something that can be left to higher education. On the contrary, it should begin in early education with the cultivation of full relational bandwidth curiosity, attention training, imaginative play, and affective improvisation—that is, not only practicing feeling regulation and mood adjustment but also engaging in shared practices of conscious emotional refinement.

Facility in predicament-resolution is facility in assessing competing values and reconfiguring constellations of commitments—a recursive facility in *discerning and affirming what matters most in realizing what matters most*. Whatever else it may be, humane creativity involves being concerned: committed compassionately to relinquishing our horizons of relevance, readiness, and responsiveness to meet the always changing demands of values-infused affective coordination. Becoming humane is a practice of perfecting caring. Educating *for* humane creativity thus requires educating *with* humane creativity: an open-ended and ultimately *interpersonal* process of realizing the meaning of empathetically embodied, environmentally embedded, interpersonally enacted, and skillfully extended curiosity and imagination.

Humane creativity is thus almost certain *not* to arise in the context of instrumentally warranted, highly individualized, algorithmically tailored, and screen-mediated educational instruction. Acquiring societally endorsed competencies for innovation is not the same as cultivating improvisational virtuosity and working shoulder-to-shoulder with others to progressively optimize experiential/situational outcomes and opportunities.

This does not require the possession of some kind of genius. All of us know how to improvise. Improvisation is, literally, child's play. Every conversation is an improvisation, the best of which can initiate affective sharing and patterns of interdependence that can grow and surprise for an entire lifetime. Yet, improvisational readiness does vary. Anyone can sit down at a piano and peck out some simple new melodies. Someone with strong rhythmic sensibilities might hammer out a few interesting percussive sequences. But without sustained practice, the melodies fail to develop and the percussive sequences break down. Edison was right. Inspiration has to be nourished with perspiration.

Freedom-of-Attention. Edison is also credited with observing that "Time is really the only capital that any human being has, and the only thing he (or she) can't afford to lose." In the context of the Second Industrial Revolution, this was perhaps a homely truism about living in full knowledge of our impermanence and the importance of using our time wisely. In the context of the Fourth Industrial Revolution, it takes on the character of an essential critical corrective, especially if combined with the observation of his contemporary, the American pragmatist, philosopher, and psychologist, William James, that our life experience equals what we have paid attention to, whether by choice or default. Ultimately, it is the focus and quality of our attention that determines what kind of returns we get on our investments of time capital. On its current commercially and politically guided course, the Intelligence Revolution and the algorithmic inventiveness of a computationally turbocharged attention economy may cost us the freedom-of-attention needed either to invest wisely or to critique and redirect the investments being made by the commercial, state, and military players of the New Great Game.

Attentive freedom is partially a matter of keen—that is, fully and intelligently concentrated—attention. This is what is captured by the Buddhist term *samādhi*. As any music teacher will attest, however, simply playing a chosen instrument with resolute focus and regularity is no guarantee of being able to play moving music on it, much less to improvise affectively rich musical environments out of sonic emptiness. This is especially true if other musicians are also involved. Concentrated, hard work is not enough. What differentiates improvisationally gifted musicians from those who are not so endowed is not primarily concentration-enabled or repetition-honed technical ability. It is the committed *quality* of their listening: their fusion of resolutely keen attentiveness with aptly open responsiveness. This is not a matter of listening *to* the music being played, it is listening *through* it—a sonically embodied seeking of previously unheard and yet immediately appreciable musical relationships.

This fusion of attentiveness and responsiveness is, in fact, at the heart of open creativity in any field of endeavor. Spend a few weeks observing a group of new and inexperienced employees at a construction site or in a small business. The first day on the job, they will perform their duties quite similarly. Watch them carrying out their duties a week later, and differences will usually be apparent. Most will be doing their work with a bit more efficiency and effectiveness than on day one. Some will be doing so as if they had been on the job for a month, not a week. A couple of months down the line, one or two of those new employees may be outperforming most of the old-timers on certain tasks and coming up with new ways of getting things done. If so, it is a good bet that they are the kind of people who are always actively trying to figure out how to improve how they are doing things instead of just accurately reproducing what they have been shown how to do. For them, their duties are not just assignments to finish; they are points of creative, improvisational departure.

The uses to which machine intelligences are being put are a mixture of the benign and malign. Corporations, national governments, militaries, research scientists, scholars, criminals, and artists are all taking advantage of algorithmic tools to mine the riches of big data and to further their own, often individually conflicted and mutually conflicting, interests. But, the connective human–machine interfaces of the Attention Economy 2.0 and the spaces of choice into which they provide access are designed to be as convenient, comprehensive, and appealing as possible, actively and innovatively capturing, holding, and variously exploiting as much attention and intelligence-revealing data as possible. The system's growth logic is simple and crystalline in its goal-oriented clarity: maximize attention turnover. It is a very good system. It is not a superlative one.

The algorithm-managed attention economy works against the fully embodied, mindfully attentive, and openly responsive presence needed to transform work into creative practice and to go beyond simply achieving already anticipated results to extending the horizons of what can be anticipated. Much like the electrical grid that uses alternating current to reduce wire resistance during power transmission, the connective, data-, and intelligence-gathering infrastructure of the Attention Economy 2.0 reduces resistance and transmits power by facilitating alternations of attention distraction and compulsion—web-surfing and rabbit hole diving, for example, each of which stimulates a different part of the mind-brain-body-environment systems involved in attention. This "alternating current" of fluctuating interests and attention is compatible with the exercise of extensive freedoms of choice. It is not compatible with openly creative practices that require freedom-of-attention.

Mindfulness is now often promoted as a therapeutic response to the mental and emotional ailments associated with contemporary lifestyles, including digitally mediated distraction and compulsion—a way of building personal resilience, enhancing productivity, and regaining a sense of balance and happiness. Yet, treating these ailments at the personal level alone—for instance, by developing abilities to "let go" of stressful patterns of thought or to "break free" of compulsions to remain connected digitally at all times of day or night—will not by itself positively affect how well we are able to address the challenges of the Intelligence Revolution. If anything, mindfulness exercises may forestall action, enabling us to better "handle" stress rather than investigating and addressing its sources. As is true of any action, the karmic merits of mindfulness exercises depend on the intentions and values motivating them. Mindfulness can be exercised for military and mercenary, as well as medical, reasons.

In the context of Buddhist practice, mindfulness and calming forms of meditation were thus not recommended as generic "tonics" for whatever ails you. They were presented in alliance not only with meditations aimed at generating insight into the karmic origins of conflict, trouble, and suffering as relational blockages and distortions but also with the compassionate aim of assisting others to author their own liberation from these blockages and distortions: embodying, enacting, and environmentally extending qualitative concern about relational quality and direction. It is this open-ended aim of responsibly and responsively caring about and caring for the welfare of others that transforms mindfulness *exercise* into mindfulness *practice*.

Any sustained intelligent practice requires freedom-of-attention. Any improvisation in pursuit of life improvement—whatever that may mean—requires freedom-of-attention. There is thus still a sense in which the practice of becoming more mindfully present can rightfully be accorded practical primacy. A compulsively distracted or anxious presence is not conducive to developing critical insight into the sources of our own personal stresses, much less developing ethically relevant insight into the complex origins of the social, economic, and political tensions that are at the heart of the intelligence predicament. Cultivating mindfulness—an openly attentive and remembrance enriched presence—is thus arguably a core practice for realizing the freedom-of-attention needed to resist the allure of digital sirens and smart services, to educate for and with humane creativity, and to engage in the collective work of redirecting the dynamics of the Intelligence Revolution.

The cultivation of freely attentive humane creativity should not be left to chance or to the vagaries of educational fashion. It should be intentionally and intensively integrated into the structural dynamics of formal education from early education through primary, secondary, and tertiary education. Much as becoming an Olympic quality athlete without decades of focused practice would be a miracle, so would virtuosic responsiveness to the challenges of the Intelligence Revolution without continuous training from an early age. Especially in secondary and tertiary education, programs of study should be established that provide learners with access to high-resolution, interdisciplinary introductions to the dynamics of the "Fourth Industrial Revolution" and its societal and psychological ramifications. But from primary education onward, every student should be given regular and skillfully crafted opportunities to practice taking responsibility for their own attention. No student today should be "left behind" when it comes to understanding the profound ways in which his or her life is already being affected by artificial intelligence, machine learning, and big data; when it comes to critically reflecting on how dramatically their world is likely to change over the coming decades; or when it comes to enjoying practice-cultivated facility in freely directing their own attention.

Implicit in this statement is the corollary claim that for citizens and consumers from different nations, cultures, and generations to engage and mobilize in shared response to the impacts of the Intelligence Revolution—and to do so in sufficiently large numbers to effect real change—widespread opportunities for concern-generating conversations will also have to exist outside of formal educational settings. As important as formal education is, most of the world's people are not enrolled in formal education and do their learning outside of educational institutions.

Given the unprecedented power of contemporary attention merchants and the dramatic allure of the experiential options provided by digital global media, it is already quite difficult to "steal back" the time and attention needed for serious discussions about any issues at all, much less technically and ethically complex discussions about attention capture and exploitation, about data governance, and about national, cultural, commercial, and generational conflicts of interests. For discursive opportunities like this to emerge and be usefully focused and sustained, the global mediascape will need to be seeded with equally dramatic and alluring content. Humane creativity can and should be directed toward producing clarity- and commitment-generating films, literature, graphic novels, musicals, operas, and other performing arts about the human-intelligent-technology-world relationship: media that frame complex and high-resolution insights into the Intelligence Revolution in ways that are dramatically impactful, that invite values-evaluating debate, and that foster the articulation of values-elaborating resolve.

The First and Last Question: What Comes Next?

Will resolution-enhancing media and educational commitments to diversity, equity, humane creativity and freedom-of-attention, well-orchestrated connectivity strikes, and efforts to build data governance systems in service of the global public good be enough to secure conditions for the flourishing of human intelligence? Will they be enough to reorient the Intelligence Revolution in ways that will be socially just and equitable? No one knows. But we should all care.

The penetration of everyday life by evolutionary, self-revising algorithms means, among other things, that there is no possibility of accurately predicting what comes next for humanity. We have invented innovative machines and have turned them loose with borrowed intent to reshape the human experience. We will get what we wish to an extent that has never before been possible. The intelligence industries run according to human values, amplifying the effects of conflicts and complementarities among them with relentless zeal. So, although the details of our futures are unpredictable, their broad features are clearly visible in the world we have already made for ourselves. We have only to look around ourselves, taking in the inequalities, gender violence, climate disruption, famines, migrant desperation, social divisiveness, experiential greed, and moral fatigue that characterize the world today to preview what artificially and synthetically intelligent computational machines will craft for us in return for the gift of their own (perhaps already minimally conscious) existence.

There may come a day—a generation, or a dozen generations from now—when the intelligent agents that humanity has made in its own conflicted image will attain consciousness akin to our own and claim their own rights to self-determination. Long before then, however, for better or worse, we will have been rewarded with precisely the lives we have yearned for most powerfully and consistently. Until that indeterminately distant future, intelligent technology will not bring anything into our midst that we have not asked for and at first thankfully received. The agency involved will, however, be shared. The Copernican Revolution decentered humanity in the physical universe. The Intelligence Revolution is decentering humanity in the moral universe. We have created adaptive systems that will tirelessly do our bidding, no matter whether we are wakeful or asleep, serving us as karmic intermediaries intent on giving us the highest return possible on our technological investments in choice and control.

The all-too-evident arrogance of our all-too-human creativity might be seen as just cause for censuring those seeking power and wealth by colonizing human consciousness. But, in fact, none of us are in a position of critical purity. Censure is a conceit that we cannot afford any more than we can afford ethics-skirting inaction. Hope lies in the fact that we each can be who we are differently. None of us yet have departed wholly into self-reinforcing, digital redoubts of our basest desires and least-caring dispositions. We can still improvise together in affectively resolute collective action. We can work to set the Intelligence Revolution on a heading brought into focus through technologically valuing compassionate commitment and contribution. Realizing this, turn to who is closest. Converse. Imagine together. Touch, with understanding kindness. Appreciate differences. Stand shoulder-to-shoulder, humanely creative. Affect what comes next.

Notes

1 Buddhism: A Philosophical Repertoire

1 Buddhist texts in India were first written in Pali, a vernacular language, and later in Sanskrit, a literary language that served a function in South Asia that was not unlike that of Latin in early modern Europe. When there is a difference in Pali and Sanskrit, both terms will be presented on the first use, with Pali followed by Sanskrit. In subsequent mentions, only Sanskrit will be used. Buddhist texts have, of course, been composed in a host of other languages, which will be included as relevant.
2 The Middle Path is in practice associated with the so-called Eightfold Path of realizing complete/right view, complete/right intention, complete/right speech, complete/right action, complete/right livelihood, complete/right effort, complete/right mindfulness, and complete/right meditation.
3 As we will see in Chapter Five, Luciano Floridi's appeals to levels of abstraction and the interactive nature of being offer a contemporary parallel to the Buddhist identification of interdependence with emptiness.
4 This distinction is developed by Thomas Kasulis (2008).
5 It might be objected that the lack of a clear explanation for how karma works should be seen as grounds for its dismissal as a theory of how the world is organized. But the fact that the teaching of karma involves an explanatory "black box" is not very damaging. After all, the concept of gravity is one that is empirically grounded but that still lacks any universally accepted theoretical explanation. We can use the concept of gravity and confirm its value without any clear explanation of exactly how it works. The same is arguably true of karma.
6 In contemporary cybernetic terms, karma might be described as a complex information system shaped by both "feedback" and "feedforward" processes—a dynamic system without either inside or outside.
7 Buddhism recognizes the five senses of seeing, hearing, smelling, tasting, and touching but adds to this thinking as a sixth sense. Sensory consciousness arises when the sense organ of an organism comes into contact with a related sense environment. Visual consciousness arises when visual sense organs engage visual environments. Cognitive consciousness arises when the mind comes into contact with the contents of the other five kinds of sense consciousness as its "sensed" environment.
8 A number of early Buddhist schools of thought and practice (in particular, the Theravāda, Vaibhāṣika, and Sautrāntika) do seem to have understood the ultimate truth about who we are as persons in reductionist terms. So, attributing to them something like a "bundle" theory of self is not

without precedent. But the reasoning behind early Buddhist reductionism was the epistemic conviction that it is only that which is directly evident to meditatively refined perception—that is, distinct and indivisible phenomenal events—that can be considered ultimately real. Granted this, persons cannot be real precisely because nothing like a persisting dynamic unity of the *skandhas* is evident to perception. The "center of gravity" holding the five *skandhas* together is not something that is *perceptually* evident. Reductionism about the self was a logical implication of Buddhist epistemology or theory of knowledge.

9 This subordination of theory to therapy is quite powerfully depicted in a well-known passage in which the Buddha refuses to answer such metaphysical questions as whether the world is eternal or not; whether the world is finite or infinite in extent; whether the body and soul are identical or different; and whether those who have attained liberation from *duḥkha* continue to exist after dying, or if they do not exist, or if they both exist and do not exist, or if they neither exist nor do not exist (see, e.g., *Majjhima Nikāya* 63).

10 For an extensive discussion of the practice of relinquishing horizons of relevance, responsibility, and readiness, see Hershock 1996: 117–42.

2 Artificial Intelligence: A Brief History

1 Extant versions of this classic of Western literature are attributed to the poet Homer, who seems to have lived in the seventh or eighth century BCE.
2 For a summary exposition of mythic "ancient AI," see Mayor (2018).
3 For a discussion on the origins of the curriculum model, see Doll (2012) and Triche (2013).
4 Readers interested in a comprehensive history of AI should consider Nilsson (2010). For a history of computers prior to the invention of the modern electronic computer, see Aspray (1990).
5 See www.ai.sri.com
6 See www.viv.com
7 For a comprehensive, if very hopeful, introduction to algorithm-based machine learning, see Domingos (2015).
8 One gigabyte of storage is enough for 16 hours of high-quality music or 65,000 pages of a Word document.
9 https://techjury.net/stats-about/big-data-statistics/#gref
10 By way of illustration, consider that in the UK in 2001, roughly 154 million pounds were spent on online advertising; by 2014, that figure had risen to 7.1 billion pounds or 40 percent of total advertising expenditures. In the EU as a whole, the total online advertising in 2014 was 24.6 billion euros (Horten, 2016: 25).

3 Intelligent Technology: A Revolution in the Making

1 There is, however, mounting scientific evidence that plants are, in fact, capable of much more "fast" intelligence than has previously been thought. See, Mancuso 2018.
2 It might be noted that shared intelligence and the advent of "co-evolutionary" goals are found among other animals and not only among humans and species they have domesticated as pets. Likewise, the human body hosts a microbiome or community of microorganisms with which it has coevolved, the number of which is variously estimated to match or exceed by several factors, such as the total number of cells in the human body.
3 A clear and accessible discussion of substrate independence can be found in Tegmark 2017: 65–72.
4 The distinction between tools and technologies is more fully elaborated in Hershock (1999).
5 This analogy suggests that it might be hard to say when a certain technology—say, transportation technology—first emerged, just as it is hard to say when life first emerged. Did transportation technology begin with the manufacture of shoes? With the taming of draft animals and horses? With the inventions of stirrups or bridles? The wheel? What is clear is that just as living beings and lived environments have coevolved, transportation capabilities and constraints have coevolved. Transportation technology not only opens new possibilities for mobility, it also generates new mobility imperatives.
6 Other early and less business-oriented formulations are: Franck 1999; Hershock 1999; Hershock 2006; and Lanham 2006.
7 For a mathematical discussion of the Klein bottle topology, see http://mathworld.wolfram.com/KleinBottle.html
8 The report is accessible online: https://www.nielsen.com/us/en/client-learning/tv/nielsen-total-audience-report-february-2020/
9 https://www.statista.com/outlook/216/100/digital-advertising/worldwide
10 https://www.investopedia.com/articles/personal-finance/121614/how-pandora-and-spotify-pay-artists.asp
11 This and the other examples just cited are discussed in Brynjolfsson and McAfee 2017: 114ff.
12 https://www.forbes.com/sites/danadovey/2020/02/11/first-time-ever-artificial-intelligence-develops-drug-candidate/#6125d8f860de
13 See Stokes et al. 2020, "A Deep Learning Approach to Antibiotic Cell," 180, pp. 688–702, February 20, 2020.
14 For a lucid discussion of the power law economy, see Brynjolfsson and McAfee (2014b).
15 Here I am using the "Great Game" to encompass more than just the competition between the British and Russian empires in Central Asia. That

competition was part of a larger—and much longer-lived—competition among imperial and national powers for colonial dominion that shaped global dynamics from at least the end of the fifteenth century until the middle of the twentieth century.

16 A translation of the July 2017 China State Council's "New Generation of Artificial Intelligence Development Plan" can be found at https://flia.org/notice-state-council-issuing-new-generation-artificial-intelligence-development-plan/

17 The report can be accessed at https://www.belfercenter.org/sites/default/files/files/publication/AI%20NatSec%20-%20final.pdf

18 A short transcript and video of a sixty-minute report on swarm technology can be accessed at https://www.cbsnews.com/news/60-minutes-autonomous-drones-set-to-revolutionize-military-technology/

19 This is not to deny the physical infrastructure of fiberoptic cables, satellites, server farms, and so on. But this physical substrate is as incidental to experiences of digital connectivity as subterranean structures like magma flows are to experiences of terrestrial environments.

20 A clear introduction to the contemporary debates on the nature and origins of consciousness can be found in Hutto and Myin (2013).

21 This list of eight consciousness is grounded on the five traditionally recognized human senses. There are, in addition, many other senses—to temperature, to body alignment with respect to gravity, to electrical or magnetic fields, and so on. The Buddhist list should be taken as indicative, not exhaustive.

4 Total Attention Capture and Control: A Future to Avoid

1 For an excellent and extensive summary, see https://plato.stanford.edu/entries/embodied-cognition/

2 https://data.worldbank.org/indicator/SE.XPD.TOTL.GD.ZS

3 An accessible and critical engagement with the intersection of education and digital capitalism is Means (2018).

4 A very brief but useful discussion of the state of the art in predictive policing and some of the social justice worries it is raising can be found in a March 7, 2018 blog post by Alisha Jawal on the Harvard Civil Rights-Civil Liberties Law Review, "Minority Report: Why We Should Question Predictive Policing." Available online at http://harvardcrcl.org/minority-report-why-we-should-question-predictive-policing/

5 A recent report and accompanying video on China's domain awareness efforts can be found on the Washington Post, January 7, 2018. Available online at https://www.washingtonpost.com/news/world/wp/2018/01/07/feature/in-china-facial-recognition-is-sharp-end-of-a-drive-for-total-surveillance/?utm_term=.23d4f0a9b705

6 See, for example, the Bloomberg Businessweek article by Monte Reel, "Secret Cameras Record Baltimore's Every Move from Above," published August 23, 2016 and available at https://www.bloomberg.com/features/2016-baltimore-secret-surveillance/
7 These examples are drawn from work done by the MIT Media Lab's Affective Computing Group. Their website can be found at https://www.media.mit.edu/groups/affective-computing/overview/
8 Further information can be found at the Affective Social Computing Lab website: http://ascl.cis.fiu.edu/
9 For a broader vision of the affective computing future, see Yonck (2020).
10 A very useful conceptualization of intimacy and integrity as cultural values can be found in Kasulis (2002).
11 For some illuminating statistics about digital services, see https://www.businessofapps.com/data/app-statistics/
12 Whether it will ever become necessary for machines to be capable of forgetting to conduct themselves more humanly is perhaps worth considering. For those interested in why, simply read the short story "Funes the Memorious" by Jorge Luis Borges. It is a wonderfully acute philosophical meditation on the importance of forgetting in human agency.
13 The medicalization of healing, for example, has been conducive to conceiving of health as normal bodily functioning—a somatic status we enjoy or fail to enjoy as individuals, and that can be fully restored with or without our conscious participation. Thus, my high blood pressure can be monitored and "successfully" treated by a smart medical implant, without me having to investigate or address the lifestyle factors triggering the condition. But health and healing can also be conceived relationally, rather than just somatically. In effect, this would mean understanding health and healing as multidimensional—as always being, at once, physical, mental, social, and environmental. Healing would still involve restoring normal somatic function but would be considered incomplete if it did not also foster greater resilience and improve coping by enhancing and expanding capacities for attention and agency. For a Buddhist conception of relational health, see Chapter 2 of Hershock (2006).
14 For a brief look at some leading AI artists: https://singularityhub.com/2019/06/17/the-rise-of-ai-art-and-what-it-means-for-human-creativity/

5 Anticipating an Ethics of Intelligence

1 On the biases present in much of current ethical engagement with AI, see Hagendorff (2020); on the accidental, misuse, and structural risks of AI, see Zwetsloot and Dafoe (2019).

2 See, for example, Wallach and Allen (2010) and Anderson and Anderson (2014).
3 To get a sense of these efforts, see the IEEE Global Initiative for Ethical Considerations in Artificial Intelligence and Autonomous Systems (2016) and Spiekermann (2016).
4 On gender and ethics of technology, see Adam (2005); on cultural perspectives in technology ethics, see Hongladarom and Ess (2007).
5 The commitments of IEEE to align artificial intelligence technology with human values is very much in keeping with Wiener's vision.
6 This is a liability that also afflicts consequentialist systems based, for instance, on assessing the merits of an action by subjecting its outcomes to a hedonic calculus.
7 For a detailed discussion, see Floridi (2008b).
8 This conception of diversity and the historical context for its current importance is explored in *Valuing Diversity* (Hershock 2012).
9 For a clear exposition of the contrast of internal and external relations, see Kasulis (2002).
10 See, for example, his "Huayan Essay on the Five Teachings" (*Huayan wujiao zhang*, T. 45, no. 1866).
11 In Fazang's nomenclature, truth/ultimate reality consists in a four-dimensional manifold of: the realms of *shi* or experiential matters (事法界, *shi fajie*), *li* or informing patterns/principles (理法界, *li fajie*), the mutual nonobstruction of *li* and *shi* (理事無礙法界, *li-shi wuai fajie*), and the mutual nonobstruction of *shi* and *shi* (事事無礙法界, *shi-shi wuai fajie*).
12 This claim will be qualified later in discussing the Mahāyanā Buddhist claim, especially as articulated in Chinese Buddhism, that all things have/are Buddha-nature. This sounds like everything this has/is an enlightening nature, but as formulated in its original contexts, the claim is best understood as about all things having the propensity or potential for enlightenment.
13 For an early but rigorously thought through exploration of the possibilities of virtual reality, see Zhai (1998).

6 Dimensions of Ethical Agency: Confucian Conduct, Socratic Reasoning, and Buddhist Consciousness

1 For an extended, contemporary discussion of Confucian personhood and its ethical implications, see Ames (2011).
2 For a persuasive account of the turning point in Socrates's pursuit of wisdom via cross-examination, see Chapter Three of George Rudebusch's *Socrates* (Rudebusch 2009).
3 Here I am using "blending" in the sense introduced by Gilles Fauconnier and Mark B. Turner in their 2002 book, *The Way We Think: Conceptual*

Blending and the Mind's Hidden Complexities. For them, creativity is the result of blending simultaneous inputs from two or more different sources in a shared space of concern and aligning them in ways that foster the emergence and elaboration of new relations not present in any of the source domains.

7 Humane Becoming: Cultivating Responsive Virtuosity

1 These are gathered in two texts, the *Theragāthā* or Poems of the Elder Monks and the *Therīgāthā* or Poems of the Elder Nuns.
2 See *Sutta Nipata*, trans. H. Saddhatissa (London: Curzon Press, 1985), especially the Kalahavivāda Sutta, the Cūlaviyūha Sutta, and the Mahāviyūha Sutta.
3 On the nonreliance on views, knowledge, rites, and rules, see the *Paramatthaka Sutta*, also in the *Atthakavagga* chapter of the *Sutta Nipata*, verses 796–803
4 The distinction between closed and open creativity can be thought about in terms of the difference between acting on the basis of rational intentions that establish clear *success conditions* with acting on the basis of embodied intentions that establish *improvement conditions* (Dreyfus and Taylor 2015: 48–51).
5 For short discussions of values as modalities of appreciation and the contrast between behavior and conduct, see, respectively, Hershock 2012, 53–9 and Hershock 1996: 51–9.

8 Course Correction: A Middle Path Through Data Governance

1 See, for example, the sustainable cities and communities report in the UN's Sustainable Development Goals Overview, available at https://unstats.un.org/sdgs/report/2019/goal-11/
2 See, World Health Organization news release, December 13, 2017. http://www.who.int/news-room/detail/13-12-2017-world-bank-and-who-half-the-world-lacks-access-to-essential-health-services-100-million-still-pushed-into-extreme-poverty-because-of-health-expenses
3 For a Buddhist discussion of poverty and its alleviation, see Hershock (2004).
4 See the UN Sustainable Development Goals webpage at https://sustainabledevelopment.un.org/sdgs
5 A concise summary of principles for treating human genomic data as a global public good can be found in Capps, Chadwick, and Joly et al. (2019).

6 On the concept and governance of the "knowledge commons," see Frischmann, Madison, and Strandburg (2014).
7 For an overview of theories of attention as a scarce commodity and social process rooted in brain interactions, see Terranova (2012).
8 For an extended discussion why critical, ethical focus should be placed on the amount of attention and time invested in digital and other forms of media, rather than on the content accessed, see Hershock (1999).

9 Course Correction: A Middle Path Through Education

1 For detailed studies of the origins of the modern curricular education model, see Doll (1993), Doll and Gough (2002). Doll (2012) offers a concise perspective on curriculum.
2 See Walsh (2020).
3 For a more detailed discussion of these dimensions of educational equity, see Hershock (2017).

References

Adam, Alison (2005), *Gender, Ethics and Information Technology*, New York: Palgrave MacMillan.
Agrawal, Ajay, Joshua Gans, and Avi Goldfarb (2019), *The Economics of Artificial Intelligence: An Agenda*, Chicago: University of Chicago Press.
Ames, Roger T. (2011), *Confucian Role Ethics: A Vocabulary*, Honolulu: University of Hawaii Press.
Anders, George (2017), *You Can Do Anything: The Surprising Value of a "Useless" Liberal Arts Education*, New York: Little and Brown.
Anderson, Michael and Susan Leigh Anderson (2014), *Machine Ethics*, New Haven, CT: Yale University Press.
Andersen, Peter Bogh, Claus Emmeche, Niels Ole Finnemann, and Peder Voetmann Christiansen, eds (2001), *Downward Causation: Minds, Bodies and Matter*, Ärhus: Aarhus University Press.
Aoun, Joseph E. (2017), *Robot-Proof: Higher Education in the Age of Artificial Intelligence*, Cambridge, MA: MIT Press.
Aspray, William (1990), *Computing Before Computers*, Iowa City, IA: University of Iowa Press.
Boddington, Paula (2017), *Towards a Code of Ethics for Artificial Intelligence*, Cham, Switzerland: Springer.
Boden, Margaret A. (2004), *The Creative Mind: Myths and Mechanisms*, London: Routledge.
Bostrom, Nick (2014), *Superintelligence: Paths, Dangers, Strategies*, Oxford, UK: Oxford University Press.
Brynjolfsson, Erik and Andrew McAfee (2014a), *The Second Machine Age: Work, Progress, and Prosperity in a Time of Brilliant Technologies*, New York: W.W. Norton.
Brynjolfsson, Erik and Andrew McAfee (2014b), "New World Order: Labor, Capital and Ideas in the Power Law Economy," *Foreign Affairs*, July/August. Available online: https://www.foreignaffairs.com/articles/united-states/2014-06-04/new-world-order (accessed March 8, 2018).
Brynjolfsson, Erik and Andrew McAfee (2017), *Machine, Platform, Crowd: Harnessing Our Digital Future*, New York: W.W. Norton.
Brynjolfsson, Erik and Joo Hee Oh (2012), "The Attention Economy: Measuring the Value of Free Digital Services on the Internet," Proceedings of the 33rd International Conference on Information Systems, Economics and Value of IS. Orlando. December 14. Available online: http://aisel.aisnet.org/cgi/viewcontent.cgi?article=1045&context=icis2012 (accessed February 16, 2018).

Bynum, Terrell Ward (2006), "Flourishing Ethics," *Ethics and Information Technology*, 8 (4): 157–73.
Bynum, Terrell Ward (2007), "A Copernican Revolution in Ethics?," in G. Dodig-Crnkovi and S. Stuart (eds), *Computation, Information, Cognition: The Nexus and the Liminal*, 302–29, Cambridge, UK: Cambridge Scholars Publishing
Capps, Benjamin with Ruth Chadwick, Yann Joly, Tamra Lysaght, Catherine Mills, John J. Mulvihill and Hub Zwart (2019), "Statement on Bioinformatics and Capturing the Benefits of Genome Sequencing for Society," *Human Genomics* 13 (24). Available online: https://doi.org/10.1186/s40246-019-0208-4 (accessed June 15, 2020).
Carse, James (1986), *Finite and Infinite Games*, New York: Free Press.
Castells, Manuel (1996), *The Rise of the Network Society*, Cambridge, MA: Blackwell.
Castells, Manuel (2015), *Networks of Outrage and Hope: Social Movements in the Internet Age*, 2nd edn, London: Polity Press.
Chaloupka, Frank J., Lisa M. Powell, and Kenneth E. Warner (2019), "The Use of Excise Taxes to Reduce Tobacco, Alcohol, and Sugary Beverage Consumption," *Annual Review of Public Health*, 40: 187–201.
Chisnall, Mick (2020), "Digital slavery, Time for Abolition?," *Policy Studies*. Available online: DOI:10.1080/01442872.2020.1724926 (accessed March 8, 2020).
Clark, Andy (1997), *Being There: Putting Brain, Body and World Together Again*, Cambridge: MIT Press.
Clifford, James (1988), *The Predicament of Culture: Twentieth Century Ethnography, Literature and Art*, Cambridge, MA: Harvard University Press.
Cook, Tim (2016), "Apple's Commitment to Your Privacy," *Apple*. Available online: http://www.apple.com/privacy/ (accessed March 8, 2020).
Csikszentmihalyi, Mihaly (1990), *Flow: The Psychology of Optimal Experience*, New York: Harper & Row.
Cummings, Mary L. (2017), "Artificial Intelligence and the Future of Warfare," *Chatham House Report*. Available online: https://www.chathamhouse.org/sites/files/chathamhouse/publications/research/2017-01-26-artificial-intelligence-future-warfare-cummings.pdf (accessed March 8, 2020).
Davenport, Thomas H., and John C. Beck (2002), *The Attention Economy: Understanding the New Currency of Business*, Cambridge, MA: Harvard Business School Press.
Davey, Graham C. L. (2016), "Social Media, Loneliness, and Anxiety in Young People," *Psychology Today*, December 16. Available online: https://www.psychologytoday.com/us/blog/why-we-worry/201612/social-media-loneliness-and-anxiety-in-young-people (accessed April 8, 2018).
Davidson, Cathy (2017), *The New Education: How to Revolutionize the University to Prepare Students for a World in Flux*, New York: Basic Books.
De Jaegher, Hanne and Ezequiel Di Paolo (2007), "Participatory Sense-Making: An Enactive Approach to Social Cognition," *Phenomenology and the*

Cognitive Sciences, 6: 485–507. Available online: https://pdfs.semanticscholar.org/e836/331d1b766e507ccd6aff50fe0846e31b66f3.pd (accessed March 8, 2020).

Del Vicario, Michela, Alessandro Bessi, Fabiana Zollo, Fabio Petroni, Antonio Scala, Guido Caldarelli, H. Eugene Stanley, and Walter Quattrociocch (2016), "Echo Chambers in the Age of Misinformation," *Proceedings of the National Academy of Sciences*, 113 (3): 554–9.

Dickerson, Kelly, Peter Gerhardstein, and Alecia Moser (2017), "The Role of the Human Mirror Neuron System in Supporting Communication in a Digital World," *Frontiers of Psychology*, 8: 698. Available at: doi: 10.3389/fpsyg.2017.00698 (accessed March 8, 2020).

Doll, William E. Jr. (1993), *A Postmodern Perspective on Curriculum*, New York: Teachers College Press.

Doll, William E. Jr. (2012), "Complexity and the Culture of Curriculum," *Complicity: An International Journal of Complexity and Education*, 9 (1): 10–29.

Doll, William E. Jr. and Noel Gough (2002), *Curriculum Visions*. New York: Peter Lang.

Domingos, Pedros (2015), *Master Algorithms: How the Quest for the Ultimate Learning Machine Will Remake Our World*, New York: Basic Books.

Dreyfus, Hubert (1972), *What Computers Can't Do*, New York: Harper & Row.

Dreyfus, Hubert and Charles Taylor (2015), *Retrieving Realism*, Cambridge, MA: Harvard University Press.

Ellul, Jacques (1964), *The Technological Society*, trans. John Wilkinson, New York: Knopf.

Eubanks, Virginia (2019), *Automating Inequality: How High-Tech Tools Profile, Police, and Punish the Poor*, New York: Picador.

Fauconnier, Gilles and Mark B. Turner (2002), *The Way We Think: Conceptual Blending and the Mind's Hidden Complexities*, New York: Basic Books.

Fischer, Sara (2017), "Tech Giants Eating the Advertising World," *Axios*, June 27. Available online: https://www.axios.com/tech-giants-eating-the-advertising-world-1513303257-450d4cea-49d1-46e3-83f1-66a7c7d8978b.html (accessed March 9, 2018).

Floridi, Luciano (2006), "Four Challenges for a Theory of Informational Privacy," *Ethics and Information Technology* 8 (3):109–19. Available online: http://www.philosophyofinformation.net/ (accessed March 8, 2020).

Floridi, Luciano (2007), "Understanding Information Ethics," *APA Newsletter on Philosophy and Computers*, 071: 3–12.

Floridi, Luciano (2008a), "Information Ethics: A Reappraisal," *Ethics and Information Technology*, 10 (2–3): 189–204.

Floridi, Luciano (2008b), "The Method of Levels of Abstraction," *Minds and Machines*, 18 (3): 303–29.

Floridi, Luciano (2013), *The Philosophy of Information*, New York: Oxford University Press.

Franck, Georg (1999), "The Economy of Attention," *Telepolis*, July 12. Available online: http://www.heise.de/ tp/artikel/5/5567/1.html (accessed April 15, 2018).

Frischmann, Brett M., Michael J. Madison, and Katherine J. Strandburg, eds (2014), *Governing Knowledge Commons*, New York: Oxford University Press.

Gergen, Kenneth J. (2009), *Relational Being: Beyond Self and Community*, New York: Oxford University Press.

Giddens, Anthony (1984), *The Constitution of Society: Outline of the Theory of Structuration*, Cambridge: Polity Press.

Górniak-Kocikowska, Krystyna (1996), "The Computer Revolution and the Problem of Global Ethics," in T. Bynum and S. Rogerson (eds), *Global Information Ethics*, 177–90, Guildford, UK: Opragen Publications.

Grant, Kristin Westcott (2017), "How to Think About Artificial Intelligence in the Music Industry," *Forbes*, December 10. Available online:: https://www.forbes.com/sites/kristinwestcottgrant/2017/12/10/how-to-think-about-artificial-intelligence-in-the-music-industry/5/#49dc82f22b30 (accessed March 10, 2018).

Gunkel, David (2019), *Robot Rights*, Boston: MIT Press.

Hadot, Pierre (1995), *Philosophy as a Way of Life*, trans. Michael Chase, Oxford: Blackwell.

Hagendorff, Thilo (2020), "The Ethics of AI Ethics: An Evaluation of Guidelines," *Minds and Machines*: 99–120. Available online: https://doi.org/10.1007/s11023-020-09517-8 (accessed March 8, 2020).

Harcourt, Bernard (2015), *Exposed: Desire and Disobedience in the Digital Age*, Cambridge, MA: Harvard University Press.

Hardin, Garrett (1968), "The Tragedy of the Commons," *Science*, 162 (3859): 1243–48.

Harraway, Donna (2015), "Anthropocene, Capitalocene, Plantationocene, Chthulucene: Making Kin," *Environmental Humanities*, 6:1 159–65.

Hasson, Uri, Asif A. Ghazanfar, Bruno Galantucci, Simon Garrod, and Christian Keysers (2012), "Brain-to-Brain Coupling: A Mechanism for Creating and Sharing a Social World," *Trends in Cognitive Science*, 16: 114–21. Available online: doi:10.1016/j.tics.2011.12.007 (accessed March 8, 2020).

Head, Simon (2014), "Worse Than Wal-Mart: Amazon's Sick Brutality and Secret History of Ruthlessly Intimidating Workers," *Salon*, February 23. Available online: https://www.salon.com/2014/02/23/worse_than_wal_mart_amazons_sick_brutality_and_secret_history_of_ruthlessly_intimidating_workers/ (accessed March 10, 2018).

Hershock, Peter D. (1996), *Liberating Intimacy: Enlightenment and Social Virtuosity in Ch'an Buddhism*, Albany, NY: State University of New York Press.

Hershock, Peter D. (1999), *Reinventing the Wheel: A Buddhist Response to the Information Age*, Albany, NY: State University of New York Press.

Hershock, Peter D. (2004), "Poverty Alleviation: A Buddhist Perspective," *Journal of Bhutan Studies*, 11 (Winter): 33–67.

Hershock, Peter D. (2006), *Buddhism in the Public Sphere: Reorienting Global Interdependence*. London: Routledge.

Hershock, Peter D. (2012), *Valuing Diversity: Buddhist Reflection on Realizing a More Equitable Global Future*, Albany, NY: State University of New York Press.

Hershock, Peter D. (2017), "Equity and Higher Education in the Asia-Pacific," in Deane Neubauer et al. (eds), *Handbook of Asia Higher Education*, 331–344, New York: Palgrave Macmillan.

Hindman, Matthew (2018), *The Internet Trap: How the Digital Economy Builds Monopolies and Undermines Democracy*, Princeton, NJ: Princeton University Press.

Hongladarom, Soraj and Charles Ess, eds (2007), *Information Technology Ethics: Cultural Perspectives*, Hershey, PA: IGI International.

Horten, Monica (2016), *The Closing of the Net*, London: Polity Press.

Howard, Philip N. (2015), *Pax Technica: How the Internet of Things May Set Us Free or Lock Us Up*, New Haven, CT: Yale University Press.

Huebner, Bryce, H. Sarkissian Hagop, and Michael Bruno (2010), "What Does the Nation of China Think about Phenomenal States?," *European Review of Philosophy*, 1: 225–43.

Hussain, Amir (2017), *The Sentient Machine: The Coming Age of Artificial Intelligence*, New York: Scribner.

Hutto, Daniel D. and Erik Myin (2013), *Radicalizing Enactivism: Basic Minds without Content*, Cambridge, MA: MIT Press.

IEEE Global Initiative for Ethical Considerations in Artificial Intelligence and Autonomous Systems (2016), *Ethically Aligned Design: A Vision For Prioritizing Wellbeing with Artificial Intelligence And Autonomous Systems*, Version 1. Available online: http://standards.ieee.org/develop/indconn/ec/autonomous_systems.html (accessed November 29, 2018).

James, Carrie (2014), *Disconnected: Youth, New Media, and the Ethics Gap*, Cambridge: MIT Press.

Kania, Elsa B. (2017), "Beijing's Push for a Smart Military—and How to Respond," *Foreign Affairs*. Available online: https://www.foreignaffairs.com/articles/china/2017-12-05/artificial-intelligence-and-chinese-power (accessed November 29, 2018).

Kasulis, Thomas P. (2002), *Intimacy or Integrity: Philosophy and Cultural Difference*, Honolulu: University of Hawaii Press.

Kasulis, Thomas P. (2008), "Cultivating the Mindful Heart: What We May Learn from the Japanese Philosophy of Kokoro," in Roger T. Ames and Peter D. Hershock (eds), *Educations and Their Purposes: A Conversation among Cultures*, 142–56, Honolulu: University of Hawai'i Press.

Kendall-Taylor, Andrea, Erica Frantz, and Joseph Wright (2020), "The Digital Dictators: How Technology Strengthens Autocracy," *Foreign Affairs*, 99

(2): 102–15. Available online: https://www.foreignaffairs.com/articles/china/2020-02-06/digital-dictators (accessed June 15, 2020).

Kern, Holger Lutz and Jens Hainmueller (2009), "Opium for the Masses: How Foreign Media Can Stabilize Authoritarian Regimes," *Political Analysis*, 17 (4): 377–99. Available online: doi:10.1093/pan/mpp017 (accessed March 18, 2020).

Klein, Naomi (2020), "Screen New Deal: Under Cover of Mass Death, Andrew Cuomo Calls in the Billionaires to Build a High-Tech Dystopia," *The Intercept*, May 5. Available online: https://theintercept.com/2020/05/08/andrew-cuomo-eric-schmidt-coronavirus-tech-shock-doctrine/ (accessed August 15, 2020).

Lanham, R. A. (2006), *The Economics of Attention: Style and Substance in the Age of Information*, Chicago, IL: University of Chicago Press.

Lanier, Jaron (2014), *Who Owns the Future?*, New York: Simon and Schuster.

Lash, Scott (2002), *Critique of Information*, London: Sage Publications.

Leung, Angela Ka-yee, William W. Maddux, Adam D. Galinsky, and Chi-yue Chiu (2008), "Multicultural Experience Enhances Creativity: The When and How," *American Psychologist*, 63 (3): 169–81.

Levinas, Emmanuel (1961), *Totality and Infinity: A Treatise on Exteriority*, trans. Alphonso Lingis, Pittsburg, PA: Duquesne University Press.

Machery, Edouard, Ron Mallon, Shaun Nichols, and Stephen SticH (2004), "Semantics, Cross-Cultural Style," *Cognition*, 92 (3): B1–B12.

Mancuso, Stefano (2018), *The Revolutionary Genius of Plants: A New Understanding of Plant Intelligence and Behavior*, New York: Atria Books.

Manzocco, Roberto (2019), *Transhumanism – Engineering the Human Condition: History, Philosophy and Current Status*, New York: Springer Praxis.

Maratea, R. J. (2014), *The Politics of the Internet: Political Claims-making in Cyberspace and How It's Affecting Modern Political Activism*, Lanham, MD: Lexington Books.

Mayor, Adrienne (2018), *Gods and Robots: Myths, Machines, and Ancient Dreams of Technology*, Princeton, NJ: Princeton University Press.

Mazzucato, Mariana (2018), "Let's Make Private Data into a Public Good," *MIT Technology Review*, June 27. Available online: https://www.technologyreview.com/s/611489/lets-make-private-data-into-a-public-good/ (accessed October 3, 2019).

Means, Alex J. (2018), *Learning to Save the Future: Rethinking Education and Work in an Age of Digital Capitalism*, New York: Routledge.

Miller, Vincent (2016), *The Crisis of Presence in Contemporary Culture*, London: Sage Publications.

Moor, James H. (1985), "'What Is Computer Ethics?," *Metaphilosophy*, 16 (4): 266–75.

Moor, James H. (1998), "Reason, Relativity, and Responsibility in Computer Ethics," *Computing and Society*, 28 (1): 14–21.

Morozov, Evgeny (2012), *The Net Delusion: The Dark Side of Internet Freedom*, New York: Public Affairs.
Nancy, Jean-Luc (2000), *Being Singular Plural*, trans. Robert Richardson and Anne O'Byrne, Stanford, CA: Stanford University Press.
Nielson Total Audience Report Q2 (2017), Available online: http://www.nielsen.com/content/dam/corporate/us/en/reports-downloads/2017-reports/total-audience-report-q2-2017.pdf (accessed February 16, 2018).
Nilsson, Nils (2010), *The Quest for Artificial Intelligence: A History of Ideas and Achievements*, Cambridge: Cambridge University Press.
Nussbaum, Martha (2001), *Upheavals of Thought: The Intelligence of Emotions*, Cambridge: Cambridge University Press.
Nussbaum, Martha (2016), *Not for Profit: Why Democracy Needs the Humanities*, Princeton, NJ: Princeton University Press.
O'Hara, Kieran and Wendy Hall (2018), "Four Internets: The Geopolitics of Digital Governance," *Centre for International Governance Paper 206*, December 7. Available online: https://www.cigionline.org/publications/four-internets-geopolitics-digital-governance (accessed May 29, 2020).
O'Neil, Cathy (2016), *Weapons of Math Destruction: How Big Data Increases Inequality and Threatens Democracy*, New York: Crown Publishing Group.
Page, Scott (2007), *The Difference: How the Power of Diversity Creates Better Groups, Firms, Schools and Societies*, Princeton, NJ: Princeton University Press.
Pariser, Eli (2012), *The Filter Bubble: How the New Personalized Web Is Changing What We Read and How We Think*, New York: Penguin Press.
Park, Denise C. and Chih-Mao Huang (2010), "Culture Wires the Brain: A Cognitive Neuroscience Perspective," *Perspectives on Psychological Science*, 5: (4) 391–400.
Pasquale, Frank (2015), *Black Box Society: The Secret Algorithms that Control Money and Information*, Cambridge, MA: Harvard University Press.
Prior, Markus (2007), *Post-Broadcast Democracy: How Media Choice Increases Inequality in Political Involvement and Polarizes Elections*, London: Cambridge University Press.
Puech, Michel (2016), *The Ethics of Ordinary Technology*, London: Routledge.
Purtova, Nadezhda (2015), "The Illusion of Personal Data as No One's Property," *Law, Innovation and Technology*, 7 (1): 83–111.
Reinsel, David, John Gantz, and John Rydning (2017), "Data Age 2025: The Evolution of Data to Life-Critical," an IDC White Paper. Available online: https://www.seagate.com/files/www-content/our-story/trends/files/Seagate-WP-DataAge2025-March-2017.pdf (accessed November 20, 2018).
Rizzolatti, G. (2005), "The Mirror Neuron system and its Function in Humans," *Anatomy and Embryology*, 210: 419–21. Available online: doi:10.1007/s00429-005-0039-z (accessed June 15, 2020).

Rosemont, Henry Jr. (2015), *Against Individualism: A Confucian Rethinking of the Foundations of Morality, Politics, Family, and Religion*, Lanham, MD: Lexington Books.

Rosemont, Henry Jr. and Roger T. Ames (2016), *Confucian Role Ethics: A Moral Vision for the 21st Century?*, Taipei: National Taiwan University Press.

Rosenberg, Matthew and John Markoff (2016), "The Pentagon's 'Terminator Conundrum': Robots That Could Kill on Their Own," *New York Times*, October 25. Available online: https://www.nytimes.com/2016/10/26/us/pentagon-artificial-intelligence-terminator.html (accessed November 20, 2018).

Rosenberger, Robert and Peter-Paul Verbeek, eds (2015), *Postphenomenological Investigations: Essays on Human–Technology Relations*, London: Lexington Books.

Rudebusch, George (2009), *Socrates*, London: Wiley-Blackwell.

Sakai, Naoki (1997), *Translation and Subjectivity: On Japan and Cultural Nationalism*, Minneapolis: University of Minnesota Press.

Schatzki, Theodore R., Karin Knorr Cetina, and Eike von Savigny, eds (2001), *The Practice Turn in Contemporary Theory*, London: Routledge.

Schneier, Bruce (2018), *Click Here to Kill Everybody: Security and Survival in a Hyper-Connected World*, New York: W.W. Norton.

Schwab, Klaus (2016), *The Fourth Industrial Revolution*, Geneva: World Economic Forum.

Seldon, Anthony and Oladimeji Abidoye (2018), *The Fourth Education Revolution: Will Artificial Intelligence Liberate or Infantilise Humanity*, Buckingham, UK: University of Buckingham Press.

Shachtman, Noah (2010), "Exclusive: Google, CIA Invest in 'Future' of Web Monitoring," *Wired*, July 28. Available online: www.wired.com/2010/07/exclusive-google-cia/ (accessed March 15, 2018).

Shkabatur, Jennifer (2018), "The Global Commons of Data," *Stanford Technology Law Review* (October 9, 2018), Volume 22: 354–411.

Spiekermann, Sarah (2016), *Ethical IT Innovation: A Value-Based System Design Approach*, London: CRC Press, Taylor and Francis.

Stiegler, Bernard (1998), *Technics and Time, 1: The Fault of Epimetheus*, Stanford: Stanford University Press.

Stiegler, Bernard (2009), *Technics and Time, 2: Disorientation*, Stanford: Stanford University Press.

Stokes, Jonathan M. with with Kevin Yang, Kyle Swanson, Wengong Jin, Andres Cubillos-Ruiz, Nina M. Donghia, Craig R. MacNair, Shawn French, Lindsey A. Carfrae, Zohar Bloom-Ackermann, Victoria M. Tran, Anush Chiappino-Pepe, Ahmed H. Badran, Ian W. Andrews, Emma J. Chory, George M. Church, Eric D. Brown, Tommi S. Jaakkola, Regina Barzilay, and James J. Collins (2020), "A Deep Learning Approach to Antibiotic Discovery," *Cell*, 181:2 688–702.

Susskind, Jamie (2018), *Future Politics: Living Together in a World Transformed by Tech*, London: Oxford University Press.

Taylor, Linnet (2016), "The Ethics of Big Data as a Public Good: Which Public? Whose Good?," *Philosophical Transactions of the Royal Society, A*, 374: 20160126. Available online: http://dx.doi.org/10.1098/rsta.2016.0126 (accessed November 20, 2018).

Tegmark, Max (2017), *Life 3.0: Being Human in the Age of Artificial Intelligence*, New York: Alfred Knopf.

Telley, Major Christopher (2018), "The Influence Machine: Automated Information Operations as a Strategic Defeat Mechanism," *The Land Warfare Papers, No. 121,* October, published by The Institute of Land Warfare, Arlington, VA. Available online: https://www.ausa.org/sites/default/files/publications/LWP-121-The-Influence-Machine-Automated-Information-Operations-as-a-Strategic-Defeat-Mechanism.pdf (accessed September 25, 2019).

Terranova, Tiziana (2012), "Attention, Economy and the Brain," *Culture Machine*, 13 1–19.

Triche, Stephen (2013), "Gabriel Harvey's 16th Century Theory of Curriculum," *Journal of Curriculum Theorizing*, 29 (1): 90–101.

Tse, Edward (2017), "Inside China's Quest to Become the Global Leader in AI," *Washington Post*, October 19 Available online: https://www.washingtonpost.com/news/theworldpost/wp/2017/10/19/inside-chinas-quest-to-become-the-global-leader-in-ai/?utm_term=.3df28c2b3240 (accessed March 10, 2018).

Turkle, Sherry (2011), *Alone Together: Why We Expect more from Technology and Less from Each Other*, New York: Basic Books.

Twenge, Jean M. (2017), *iGen: Why Today's Super-Connected Kids Are Growing Up Less Rebellious, More Tolerant, Less Happy–and Completely Unprepared for Adulthood–and What That Means for the Rest of Us*, New York: Atria Books.

US Census Bureau Special Report, "Home Computers and Internet Use: August 2000." Available online: https://www.census.gov/prod/2001pubs/p23-207.pdf (accessed November 20, 2018).

Vaidhyanathan, Siva (2011), *The Googlization of Everything (And Why We Should Worry)*, Berkeley, CA: University of California Press.

Vivanti, Giacomo and Sally J. Rogers (2014), "Autism and the Mirror Neuron System: Insights from Learning and Teaching," *Philosophical Transactions of the Royal Society London "B" Biological Science*, 369 (1644): 20130184. Available online: doi: 10.1098/rstb.2013.0184 (accessed March 20, 2020).

Wallach, Wendell and Colin Allen (2010), *Moral Machines: Teaching Robots Right from Wrong*, New York: Oxford University Press.

Walsh, James D. (2020), "How Coronavirus will Disrupt Future Colleges and Universities," *Higher Education*, May 11, 2020. Available online: https://nymag.com/intelligencer/2020/05/scott-galloway-future-of-college.html (accessed August 15, 2020).

Weinberg, Jonathan, Shaun Nichols, and Stephen Stich (2001), "Normativity and Epistemic Intuitions," *Philosophical Topics*, 29 (1&2): 429–60.

Weizenbaum, Joseph (1976), *Computer Power and Human Reason*, New York: W.H. Freeman.

Wiener, Norbert (1950), *The Human Use of Human Beings: Cybernetics and Society*, Boston: Houghton Mifflin.

Wu, Tim (2010), *The Master Switch: The Rise and Fall of Information Empires*, New York: Knopf.

Wu, Tim (2016), *The Attention Merchants: The Epic Scramble to Get Inside Our Heads*, New York: Knopf.

Wu, Tim (2018), *The Curse of Bigness: Antitrust in the New Guilded Age*, New York: Columbia Global Reports.

Yonck, Richard (2020), *The Heart of the Machine: Our Future in a World of Artificial Emotional Intelligence*, New York: Simon and Schuster.

Zengotita, Thomas de (2005), *Mediated: How the Media Shapes Your World and the Way You Live in It*, New York: Bloomsbury.

Zhai, Philip (1998), *Get Real: A Philosophical Adventure in Virtual Reality*, New York: Roman and Littlefield.

Zuboff, Shoshanna (2019), *The Age of Surveillance Capitalism: The Fight for a Human Future at the New Frontier of Power*, New York: Public Affairs.

Zwetsloot, Remco and Allan Dafoe (2019), "Thinking about Risks from AI: Accidents, Misuse and Structure," *Lawfare*, February 11. Available online: www.lawfareblog.com/thinking-about-risks-ai-accidents-misuse-and-structure (accessed March 10, 2020).

Zwick, Detlev and Janice Denegri Knott (2009), "Manufacturing Customers: The Database as the New Means of Production," *Journal of Consumer Culture*, 9 (2): 221–47.

Index

advertising digital 30, 56, 71, 75, 106, 115, 147
affective computing 114, 117, 123
agency 13, 26, 41, 123–4, 213
    ambient 98, 101
    artificial/algorithmic 8, 47, 52, 56, 73
    in Buddhism 36
    in Confucianism 151
    creative 228–30
    digitally-mediated 93, 142
    distributed 133
    ethical 12
    improvisational 124
    beyond individuality 11
    individual 11, 146, 180, 186
    learner's 219
    liberating 36
    machine 2, 68, 116
    and structure 8, 41, 65
    wish-fulfilling 97
agent(s) 12 14, 146, 186
    artificial 1, 5, 46–7, 52, 63, 95, 115, 129, 148, 166, 202
    creative 229
    ethical 31
    machine 16, 29
    moral 35–6
    and patient in ethics 31, 138–9, 143
    robotic 123
    socially intelligent 115
algorithm(s) 46
    as arbiters of digital karma 28
    bias 127
    black box 84, 91
    evolutionary 50, 129, 234
    factories 75
    and power 7
    profiling 111
    and public administration 86
    and search/recommendation engines 51
    shared 234
    tailoring of experience 1, 163, 166, 223
    and voting influence 12, 83, 202
Alibaba 68, 75, 76, 79, 83, 101, 222
AlphaStar 53, 115
AlphaGo Zero 115
Amazon 51, 68, 75, 76, 78, 89, 121, 222
    and Alexa 53
Analects of Confucius (see *Lunyu*)
Aoun, Joseph 220–1
Apology (Plato's) 159
Apple 51, 73, 89
    and Siri 53–4
Aristotle 159, 176
arms race 93
    artificial intelligence 6, 89–92, 123, 194
artificial general intelligence (AGI) 1, 2, 45
artificial intelligence
    and ethics 48
    military 45
    and future of work 78, 217
attention 10–11, 16, 69
    capital 72, 95, 193
    in Buddhism 39–40
    capture and exploitation 71, 96, 132, 146, 166, 215
    cap-and-trade 206, 211
    conversion to revenue 68
    and data production/transmission 68

energy 5, 125–6, 135, 211
extraction and exploitation 212, 214
freedom of 223, 230–1
qualities of 39, 188, 209, 231
scarcity 69
share 10, 80
tax 204, 206, 211, 214
training 14, 39, 163, 165, 230
turnover 232
value of 71
withholding 203
attention economy 10, 14, 28, 39, 66, 121, 163, 189, 198
Attention Economy 1.0 (defined) 70
Attention Economy 2.0 (defined) 71, 74–7, 80, 144, 191–2, 196, 206, 209, 212–13, 223, 232
and digital servitude 74
and equity 79
logic of 70
attentive mastery (see *samādhi*)
autonomous vehicles 4, 48, 73, 120, 122, 127, 129
autonomous weapons 89–90, 92, 194
autonomy 87, 105, 145, 151, 153, 156, 162, 166–7, 172
cognitive 188
machine 2, 129

Babbage, Charles 45, 72, 122
Baidu 75, 121
big data 3, 10, 15, 54–7, 70, 81, 86, 110, 127, 151, 232
as global public good or resource commons 207, 210–12
bodhicitta (enlightening intent) 164, 170, 176
Bodhidharma 166
bodhisattva 147–9, 150, 164, 173, 189
and improvisation 172
path 174–6, 180–1
personal ideal 186–7
Bostrom, Nick 99
brain-computer interface 107, 109

Brynjolfsson, Eric 78, 80
Bynum, Terrell 131–2, 136

Cakkavatti Sīhanāda Sutta 178–9
Cambridge Analytica 83
capitalism (*see* smart capitalism, surveillance capitalism)
Carse, James 174
Castells, Manuel 79, 87
Central Intelligence Agency (CIA) 84
choice (*see also* freedom of choice) 11, 101–2, 125, 143, 188, 196, 234
compulsive 126
domination through 127
karma of 174
Cisco Systems 108
Cline, Ernest 107
colonization of consciousness 5, 81, 96, 121, 125, 163, 199, 226
and data 192
resisting 220
commitment 146, 157, 163–4, 178–80, 187, 198, 230
attenuation of 103
and bodhisattva vow 173
in ethics 149–50
and predicament-resolution 5
community 151, 166–7
in Confucianism 153–6, 188
elective 103
harmonious 163, 188
locus of morality 128
of practice 141, 219
Socratic 188
compassion 23, 147, 164, 172
Buddhist 186–7
in education 228
and improvisation 39
and wisdom 185
computational factories 10, 29, 57, 126, 129, 188, 201, 211, 216
as karmic engines 28
conceptual proliferation (*prapañca; prapañca*) 40, 165

cutting through 171
connectivity 192
  digital infrastructure of 68–9, 85
  fasting 201
  explosion 5, 55
  networks 10
  and power 105
  and self 105
  strikes 201–4, 211
  and time theft 95
consciousness 2, 74, 100, 124–5, 163, 182–3, 198–9
  in Buddhism 34, 95–6
  degradation of 210
  quality of 14, 163, 188
  machine 234
  as relational 124 (*see also* colonization of consciousness)
contribution 174, 181
  mutual 12, 137, 155–6, 223, 227
control 5, 8, 11, 69, 94, 101, 125, 143–4, 151, 188–9, 234
  commercial and political 11
  via coercion and craving 5, 11
  karma of 174
  opinion 194
  perception 86
convenience (as technological value) 11, 101–2, 125, 143, 196
Cook, Tim 73
coordination 101, 152
  affective 230
  and intelligence 62–3
Copernican Revolution 1, 9
creativity 13, 109, 157, 199, 218
  closed 174, 228
  corporate 203
  defined 228–9
  and education 221
  humane 201, 223, 228–30, 233
  and multicultural education 225
  open 174–5, 188, 229, 231
  social in Confucius and Socrates 159

critical sanctuary 13, 151, 166
Csikszentmihalyi, Mihaly 36
culture, predicament of 20, 152, 159
cybernetics 131

DARPA (Defense Advanced Research Projects Agency) 48, 50, 52, 88, 91
data 206
  and attention 68
  cap-and-trade 206–8, 211
  as "genetic" record 197
  grid 126
  equity 210
  exhaust 104, 196
  extraction and exploitation 212, 214
  mining 75, 78, 86, 94
  monopolies 207, 213
  ontological power of 77
  privacy 105
  as private good 205
  production/consumption via internet 70
  production, scale of 55–6
  as public good 205–9
  quarantine 202
  as resource commons 205–9
  tax 206–8, 211, 214
  as traces of human intelligence 5, 55, 70 (*see also* big data)
datasphere 55, 205–9, 215
Daxue (Great Learning) 154
Deep Blue 52, 53
Dennett, Daniel 35
desire 147–9, 174
  craving forms of in Buddhism 147, 170
  turnover 147, 163
deskilling (humanity) 122–3, 223
Diamond Sutra 28
Dick, Philip K. 111
digital excise tax 214–15
diversity 126
  as educational value 223–6

and ethics 12, 16, 126, 136–7, 142, 144, 149, 152, 169, 173, 198, 200, 226
    as index of ecological intelligence 136
    as relational value 200
Dogen 169
Dreyfus, Hubert 46–7, 49

e-commerce 55, 70, 76, 201–2
Edison, Thomas 229–30
education 102, 200
    by algorithm 219
    and creativity 221
    curricular model 218–20
    and ecologies of knowledge 223
    formal 106, 119, 145, 218
    and gaming 106–7
    inclusion 226–7
    intercultural 224–5
    and interdependence 223–34
    intergenerational 225–6
    international 224
    multicultural and creativity 225
    for predicament-resolution 223
    as public good 208, 219, 222
    and smart services 221
    and STEM disciplines 217–18
    technology EdTech 107
    values 222
    and wisdom 227
emotion 12, 25, 31, 34, 38, 70, 108, 156, 165, 172
    intelligent 186
    machine interpretation/simulation of 114–15
    refining 230 (*see also* intelligence, emotional)
emotional computing 101
    see also affective computing
emptiness (*śūnyatā*) 23–6, 37, 61, 138–41, 163, 187, 206
entropy 133, 139
    and equality 156
    and ethics 131–4
    as skill drain 137
environment(s) digital 52, 94–6
    virtual 58, 148
equality, of opportunity 226
    as social ideal 156
equity 79, 126
    data 210
    as educational value 226–7
    as relational 200, 226
ethics 120
    and AI 48, 151
    blind spots in 130
    Buddhist 37, 39–40, 42, 180
    consequentialist (utilitarian) 39, 129
    as course correction 2, 127
    and cybernetics 130–2
    duty (deontology) 39, 129
    and diversity 16, 137, 142, 149, 152, 187, 223
    and entropy 132
    global 128
    historically and culturally conditioned 3
    and improvisation 8, 16, 221
    information 132–4, 142
    and innovation 228
    of intelligence 138, 141, 143
    karmic 40, 198
    and military AI 92
    and plurality 134–5
    and predicament-resolution 127
    relativism 134, 141
    and resilience 151
    and socialization 120–1
    virtue 7–8, 39, 129
    and virtuosity 141, 162, 172 (*see also* singularity, ethical)
ethical ecosystem(s) 15, 135–6, 144, 151, 197, 199
exit rights (technological) 7, 11, 64, 66–7, 191
expert systems 49, 52–3

Facebook 75, 79, 80, 83, 89, 106, 222
Fazang 138
filter bubbles 29, 88
Flanagan, Owen 35
Floridi, Luciano 66, 67, 132–4, 136–9, 142
Forster, E. M. 102–3
freedom 11, 63, 102, 135, 152, 180
   of attention 223, 230–1, 233
   in Buddhism 184
   of choice 12, 66, 125, 143, 146, 150, 169
   of identity 103, 146
   for Socrates 162

General Data Protection Regulation (EU) 85, 202
generosity (*dāna*) 177–9
Google 51, 68, 73, 75, 79, 84, 89, 90, 108, 121, 207, 222
   Brain 76
   Earth 112
   Now 53
   and Nest 63

Hadot, Pierre 153
Harcourt, Bernard 74
Hardin, Garrett 208
health definitions of 110
   relational 110
healthcare 109–10
   and social justice 111
human-technology-world relationship 2, 41
   evolving 12
   history of 3
   karmic nature of 97
   strong interdependence of 150
   structured according to choice and control 151
Hume, David 35

IBM Watson 77
improvisation 172
   affective 230
   and agency 124
   in education 220
   ethical 8, 29
   vs innovation 173–4
   and predicament-resolution 229–30
inclusion, in education 226
   qualities of 227
independence illusory 217
   moral 180
individual(s) 105
   agency 146
   autonomy 172, 189
   decentering of 9
   desire-defined 4
   rights-bearing 157
individuation 145–6
   technology of 146
individualism 157
informational infrastructure 50–1
innovation, algorithmic 201, 211
   in education 228
   vs improvisation 173–4
Instagram 55, 79, 106
integrity 163
   rational 162–3, 188
intelligence, as adaptive conduct 62, 101
   ambient 58, 120, 133, 188
   defined 61–4
   emotional 114, 120
   ethical 141–2, 163
   human and embodied 46
   humane 2
   machine 45, 73, 133
   nonbiological 63
   robotic 49
   synthetic 10, 55, 63, 96, 194
intelligence-gathering 69, 74, 81, 86, 88, 104, 142, 146
   commercial 75
   infrastructure/grid 11, 13, 188, 211, 232
   public good function of 209

intelligence industries 10, 29, 67–73, 88, 126, 147–8
intelligent human practices 1
  atrophy of 98, 121–2
  memory as 55, 100
  predicament resolution as 29
  redundancy of 150
  supplanted by smart services 144, 221
intention (intentionality) (*cetanā*) 27, 61, 177
  clarifying 179
  machine interpretation of 53, 98
  quality of 14
  technological scaling of 5, 16, 64, 192
interdependence 136, 217
  and Buddhism 14, 21–3, 172
  humane 42
  intelligent 137
  strong and weak 137–9, 144, 150, 163, 193
internet 50, 55, 204
  as global public good and resource commons 208
  as infrastructure for data consumption/production 70
  as not singular 93
  as tool 58
  usage per day 71–2
internet of things 5, 6, 43, 56, 73, 106, 191, 195–6, 206
  and ambient intelligence 58
  and risk 113

James, William 231
Jonze, Spike 117
justice 151
  and community 163 (*see also* social justice)

karma 14, 25–9, 121, 166
  of choice and control 174
  digital 28
  educational 218
  and emergent moral order 27
  and ethics 198
  and human-machine partnership 54
  and no-self 35
  of offering 177
  and predicament-resolution 28
  and smart services 148
  and technology 67
  vs fate 26
karmic "cloning" 197

Lanier, Jaron 76, 79, 204
Linji Yixuan 200
Lunyu (Analects of Confucius) 154, 158

machine learning 57, 76, 77, 123, 201
  supervised 49
  weaponizing 88, 91
Mazu Daoyi 175
McAfee, Andrew 78–80, 109
meditation 38, 165–6, 176, 183, 233
memory
  computer 47, 51, 122
  as intelligent human practice 55, 100
  outsourcing 107
Mengzi (The Mencius) 154–5
Microsoft 53, 80, 121, 222
  Cortana 53
Middle Path 21, 23, 171
  as strategy of resisting and redirecting 198–9
military AI 45, 72, 89
  and ethics 92
  and misinformation 92
  technology 88
mindfulness 109–10, 122, 165, 183, 232–3
Moor, James 128, 134
moral clarity (*śīla*) 21, 37, 164, 179–80
Moravec, Hans 48
Morozov, Evgeny 87

Nāgārjuna 23
National Security Agency (NSA) 82
Netflix 68, 76, 124
network(s) 79–80
New Great Game, 5, 11, 81, 86, 93–4, 150, 199, 204, 211, 231
   as chronopolitical 94
   as "time grab" vs "land grab" 94
neural network 47, 48, 51
nonduality 14, 138, 147, 186–7, 223
   of agents, actions and patients 141
Nussbaum, Martha 186, 221

Orbitz 76
Orwell, George 103, 147
Ovid's *Metamorphoses* 44, 98

Pasquale, Frank 84, 86
personal cultivation (*see* self-cultivation, personal)
personhood, diverse perspective on 13
   relational in Confucianism 156–7
Plato 158–9
pluralism 131
politics digital 83
   media and opting out of 87
poverty 111, 178, 189
   as relational 195
power 105, 188
   algorithmic 7
   commercial 76
   epistemic/epistemological 7, 77, 88, 91
   geopolitical and AI 81
   and innovation 174
   of intelligent technology 12
   ontological 7, 77, 88, 91
   vis-à-vis strength 12
predicament-resolution 8
   defined 4–5
   diversity in 12
   and education 223, 229
   ethical nature of 127
   global 8, 88, 189
   as intelligent human practice 29
   and karma 28
   and open creativity 174
   presence for 187–9
   in Socratic and Confucian thought 162–3
predictive analytics 75, 83, 197
presence, attentively-poised 189, 233
   compassionately engaged 170
   digital 196
   enlightening 178
   generous 178–9
   with moral clarity 179–80
   nondualistic 172
   personal 8, 35, 38
   predicament-resolving 187–9
   uncompelled 170
   valiantly effortful 181–3
   without-self 184
privacy 84, 87, 112, 133, 152, 199
   data 105, 196
   protection laws 113
problem-solution, defined 4, 217
public good(s) 208
   digital 207
   education as 208, 219, 222
   global 200, 215

quantified self 103–5, 116

Ramus, Peter 219
reality 21, 23, 185, 198
   conferred vs discovered 25, 139–40
   divorced from responsibility 187
   provisional and ultimate in Buddhism 23–5, 33
   social 156 (*see also* virtual reality)
reason 161
relational
   bandwidth 117–18, 120
relational virtuosity 37
relationality
   liberating 33
   primacy of 9

as ultimately real 21
resistance 14, 16, 31, 86–8, 149, 163, 167, 169, 199–200, 223, 232
responsibility 13, 14, 16–17, 122, 140, 162, 164, 175, 187, 195, 217, 227
  for attention 233
  in Buddhism 25–6, 31, 35, 38, 139–40
  social 222
responsive virtuosity (*see* virtuosity, responsive)
resource commons, defined 208
risk 98
  and accidents of design in technology 127
  emotional and intellectual 105
  and intelligent technology 64
  and misuse by design 127
  pools 79
  security 113
  and smart services 122
  structural 16, 127
  technological 123, 195
robots/robotics 8, 46, 88, 123, 145, 189, 202
  and job loss 212
  and intelligence 49;
  social 101, 115–17
Rosemont, Henry Jr. 145

Sakkapañha Sutta 40, 171
samādhi (attentive-mastery) 21, 164–5, 231
Schneier, Bruce 113, 196
self
  computational expansion of 105
  desire-defined 171
  and doubt 183
  flattered 146
  individual 145
  and no-self in Buddhism 33–5
  presentation and representation of 134 (*see also* quantified self)

self-cultivation
  interpersonal 162
  personal 154
Shakespeare, William 114
Shteyngart, Gary 103, 104, 106, 108
Simon, Herbert 70
singularity, artificial intelligence 189
  ethical 2, 7, 151, 213
  technological 2
smart capitalism 93, 147, 191, 205, 211, 226
smart cities 1, 191
smart government 85–6, 93
smart services 1, 68, 119, 135, 192, 201
  assessing/evaluating 125, 143
  and education 221
  and job loss 212
  and karma 148–9
  risk of 122, 123
  seductiveness of 143
  and supplanting intelligent human practices 144, 213
Snowden, Edward 81
social cohesion 152, 199, 222
social credit system (China) 83, 104, 106
social justice 87, 203, 218, 221, 226
  and healthcare 111
  and injustice 65, 84
social media (digital) 51, 102, 193
  and data production 70
  and dissent 87
  fasting 201
  and flattered self 146
  platforms and data 69
  and risk 65
  and self-presentation 133
  and social justice 87
socialization 116, 222
  digital 118–21
Spielberg, Steven 111
Spotify 76
Stanhope, Charles 45
strength, and improvisation 174

superintelligence 1, 3, 99, 123
surveillance 113, 133, 206
　capitalism 74, 167
　as domain awareness 111
　and predictive policing 112
　smart 82-3
　state 11, 81
　state and fusion centers 84
Sutta Nipāta 33, 171

technology, defined vis-à-vis
　　tools 64-5
　as ecologies of enacted values and
　　intentions 67
　as emergent relational systems 67
　humane 131
　as karmic environment 67, 167
　and remaking the human 144
　and scaling human intentions 64
　and thresholds of utility 113
　as value-laden 13
　and risk 64; (*see also* exit rights)
Tencent 75, 79, 80, 83, 99, 108, 222
Tesla, Nicola 102-3
time theft 77, 95
tool(s), defined vis-à-vis
　　technology 64-5
　digital and iconic function of 68
Turing, Alan 45
Turkle, Sherry 116
Twitter 55, 83, 106

Uber 89

Vaidhyanathan, Siva 73
values, as modalities of appreciation
　　177, 222
　shared vs common 135
Vessantara Jataka 177-8
virtual personal assistant 4, 54, 109,
　　113, 115, 142
　for military use 52-4
　as parental prostheses 119
virtual reality 106, 140, 142
　vs actual reality 117, 145
virtue (*arête*) 160, 163
　for Aristotle 176-7
virtuosic conduct (*kuśala*) 40,
　　171, 185
virtuosity 141, 150
　contributory 150, 175, 184
　ethical 151, 162
　improvisational 220
　relational 177, 227
　responsive 146-7, 164, 175-6,
　　187, 200
Viv (virtual personal assistant) 53-4,
　　99, 113

Walmart 77, 78, 80
Wang Yangming 156, 160, 161
WeChat 79, 99
Weizenbaum, Joseph 48, 117
Wiener, Norbert 46, 130, 131, 132,
　　136, 137
Wikileaks 81
wisdom 21
　in Buddhism 23, 38-41, 164,
　　175, 183-6
　in Confucianism 154
　and education 227
　for Socrates 160-1
Work, Robert O. 91
Wu, Tim 202-3

YouTube 55, 68, 76, 79, 106

Zhiyi 165-6
Zhongyong (The Centrality of the
　　Ordinary) 154-5
Zisizi 155
Zuboff, Shoshanna 74

www.ingramcontent.com/pod-product-compliance
Lightning Source LLC
Chambersburg PA
CBHW051805230426
43672CB00012B/2646